Advances in Gyroscope Technologies

AF302475

M. N. Armenise · C. Ciminelli ·
F. Dell'Olio · V. M. N. Passaro

Advances in Gyroscope Technologies

Springer

M. N. Armenise
Dipartimento di Elettrotecnica
  ed Elettronica
Politecnico di Bari
Via Re David 200
70125 Bari, Italy
e-mail: armenise@poliba.it

C. Ciminelli
Dipartimento di Elettrotecnica
  ed Elettronica
Politecnico di Bari
Via Re David 200
70125 Bari, Italy

F. Dell'Olio
Dipartimento di Elettrotecnica
  ed Elettronica
Politecnico di Bari
Via Re David 200
70125 Bari, Italy

V. M. N. Passaro
Dipartimento di Elettrotecnica
  ed Elettronica
Politecnico di Bari
Via Re David 200
70125 Bari, Italy

ISBN 978-3-642-42345-1

DOI 10.1007/978-3-642-15494-2

ISBN 978-3-642-15494-2 (eBook)

Springer Heidelberg Dordrecht London New York

Cover design: eStudio Calamar S.L.

Printed on acid-free paper

Springer is part of Springer Science+Business Media (www.springer.com)

Preface

The gyroscope, which measures angular rotation around a fixed axis with respect to an inertial space, is a key sensor in modern navigation systems enabling to plan, record and control the movement of a vehicle from one place to another. This device has a wide spectrum of applications in space engineering, aeronautical and military industry, automotive, medicine and so on. For these reasons gyro architectures and technologies have been investigated by very important research groups in USA, Europe and Asia. National space agencies of a number of Countries have invested significant financial resources for developing innovative gyroscope technologies. Results of this intense research effort have been reported in a large number of scientific papers and patents.

The purpose of the book is to collect and critically review the main results obtained by the scientific community in advanced gyroscope technologies. Architectures, design techniques and fabrication technology of angular rate sensors proposed in literature are described. Future research trends aimed to cover special applications are also considered.

The book is intended for researchers and Ph.D. students interested in modelling, design and fabrication of gyros. It may be a useful education support in some university courses focused on gyro technologies.

In recent years the authors have spent an intense research effort on optical angular velocity sensors working on some specific projects supported by some space agencies. They use their deep know–how on different gyroscope technologies to offer to the readers a wide vision about the book subject.

The book includes seven chapters. First two chapters introduce the topic and briefly describe physical effects exploited in gyroscope technologies. Chapters 3, 4 and 5 are focused on optical gyros. State-of-the-art of ring laser gyros, fiber optic gyros and integrated optical gyros is accurately reviewed. Vibratory gyros and, in particular, MEMS gyros are the topic of Chap. 6, where MOEMS gyros are also introduced. Finally, the book topic is summarized in Chap. 7 that offers also an overview of the most innovative technologies for angular rate sensors with out-standing performance.

Bari, May 2010   M. N. Armenise

C. Ciminelli

F. Dell'Olio

V. M. N. Passaro

Contents

Chapter 1
Introduction

Inertial sensors, which allow to measure linear acceleration and angular velocity, are emerging as a crucial class of sensors and their applications are continuously increasing. They were first developed for aerospace and military systems but currently are extensively used in a wide spectrum of applications, such as automotive, medicine, consumer electronics and so on.

Inertial Measurement Units (IMUs) enabling the measurement of acceleration along three axes and angular velocity around the same three axes are very sophisticated systems whose development in terms of cost reduction and performance improvement is considered an important task for naval, defense and aerospace industry. Innovative space-qualified IMUs having reduced weight, low power consumption, high sensitivity, and good reliability are needed for new space missions. IMU worldwide market, which is currently around 2 billion dollars, is expected to significantly grow up in the next few years.

Gyroscope (also named gyro) measures the angular rate around a fixed axis with respect to an inertial space. In the last four decades an intense research effort has been devoted to design, optimize and fabricate different kinds of gyros essentially based on angular momentum conservation, Sagnac and Coriolis effects. In the last few years development of innovative gyros has been focused on micro-photonics and micro-electro-mechanics technologies.

In this chapter main technologies and applications of angular rate sensors are briefly discussed. Performance parameters allowing the comparison of different gyros are defined, too.

1.1 Overview of Gyroscope Technologies

It is possible to recognize three different kinds of gyroscopes: spinning mass gyros, optical gyros and vibrating gyros. In the first category all the devices having a mass spinning steadily with respect to a free movable axis fall. Optical gyroscopes are based on Sagnac effect which states that phase shift between two waves counter-propagating in a rotating ring interferometer is proportional to the loop

M. N. Armenise et al., *Advances in Gyroscope Technologies*,
DOI: 10.1007/978-3-642-15494-2_1, © Springer-Verlag Berlin Heidelberg 2010

angular velocity. Vibrating gyros are based on Coriolis effect that induces a coupling between two resonant modes of a mechanical resonator.

The basic configuration of gyroscope exploits the inertial properties of a wheel (or rotor) spinning at high speed, which tends to keep the direction of its spin axis by virtue of the tendency of a body to resist to any change in the direction of its moment. In the 1960s Dynamically Tuned Gyroscope (DTG), which is based on this physical principle, was developed [1]. DTG was used for many years in aerospace and military industry and it was included in the IMU of the Space Shuttle.

One of the most successful spinning mass gyroscope is the Control Moment Gyroscope (CMG) [2] which is widely used for satellites stabilization [3]. It consists of a spinning rotor and one or more motorized gimbals that tilt the rotor angular moment. As the rotor tilts, the changing angular moment causes a gyroscopic torque that rotates the spacecraft. CMGs were used for decades in large spacecraft, including Skylab, Mir Space Station and the International Space Station.

Miniaturization of spinning mass gyros is very difficult and their consequent decline has created interesting business opportunities for vibrating and optical gyros that can be effectively miniaturized by MEMS and integrated optical technologies, respectively.

In the 1980s, a highly performing vibrating gyro, the Hemispherical Resonator Gyro (HRG), was developed. HRG sensing element is a fused silica hemispherical shell (diameter around 30 mm) covered by a thin metal film [4]. This device is a very sensitive and expansive gyro and it was used in some space missions, including the Near Earth Asteroid Rendezvous and the Cassini ones.

Silicon and quartz MEMS gyros are innovative miniaturized vibrating angular rate sensors. They assure low cost and performance which is constantly increasing. MEMS gyros market is quickly growing up and is reaching 800 millions dollars in 2010 [5].

Since 1963 when the first Ring Laser Gyroscope (RLG) based on Sagnac effect was fabricated [6], a number of photonic gyroscopes have been proposed and demonstrated, including Fiber Optic Gyroscopes (FOGs) and integrated-optics gyroscopes [7, 8]. In the 1990s, the first FOG in space was used in X-ray Timing Explorer mission [9].

Recently, other very sophisticated technologies for future gyros have been demonstrated, including the nuclear magnetic resonance gyro [10] and the superfluid gyro [11].

Some reviews on gyro technology are reported in literature [12–15], while the most recent advances are in this book.

1.2 Gyro Performance Parameters

Different gyro technologies are usually compared in terms of cost, power consumption, reliability, weight, volume, thermal stability, immunity to external disturbance and other very important performance parameters describing the gyro behavior when it is set up in a more complex system.

Starting from gyroscope static input–output characteristic a number of gyro performance parameters can be defined such as scale factor, bias, input and output range, full range, resolution, dynamic range and dead band [16].

Gyro scale factor is defined as the ratio between the change in sensor output and the relevant angular velocity variation. It is usually evaluated as the slope of the straight line that can be obtained by linear fitting input–output data.

Bias is defined as the average, over a specified time interval, of gyro output that has no correlation with either input rotation or acceleration. Bias is measured in °/h or °/s.

Input range is the range of input values in which the gyro performance matches a specified accuracy. Output range is the product between input range and scale factor. The algebraic difference between the upper and lower values of the input range is called full range.

Minimal detectable angular rate or resolution (expressed in °/h or °/s) is the minimum angular velocity that can be detected by a gyroscope. The ratio between full range and resolution is called dynamic range (dimensionless quantity).

Finally, dead band is a range between the input limits within which variations in the input produce output changes less than 10% of those expected.

Gyro frequency response or step response allow to calculate gyroscope bandwidth and response time.

Main noise contributions in gyroscope are:

- quantization noise,
- bias instability (or bias drift),
- angle random walk.

Quantization noise is mainly due to analog-to-digital conversion of the gyro output signal whereas the cause of other noise contributions depends on the gyro operating principle.

Bias instability is the peak-to-peak amplitude of the bias long term drift. It is expressed in °/s or °/h.

Angle random walk (ARW) is a noise contribution to the rotation angle value obtainable by integrating the angular velocity. Standard deviation of noise observed in rotation angle estimation can be written as:

$$\sigma_{rw} = W_{ARW} \sqrt{t} \tag{1.1}$$

where W_{ARW} is the angle random walk coefficient, usually expressed in $°/\sqrt{h}$ or $°/\sqrt{s}$.

Different noise contributions can be estimated by modeling the noise in angular rate measurement by a stochastic process [17]. A stochastic process $u(t)$ consists of a family of time dependent real functions $u(t,r)$, each of them associated to an element r of a probability space $\Re$. The real function $u(t,r)$ (also denoted as $u(t)$) is called trajectory of the stochastic process. If we consider a stationary stochastic process, its autocorrelation is given by:

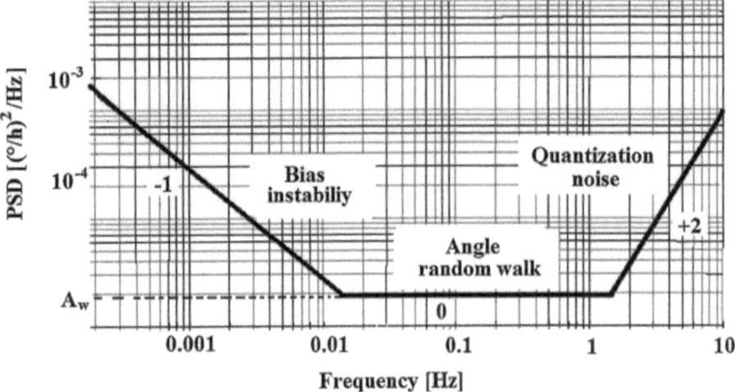

Fig. 1.1 Piecewise representation of noise PSD in the angle rate measurements

$$\phi_u(\tau) = E[u(t+\tau)u(t)] \tag{1.2}$$

where E is the expectation operator. If the stochastic process is ergodic, auto-correlation can be calculated by any trajectory $u(t)$ as:

$$\phi_u(\tau) = \int\limits_{+\infty}^{+\infty} u(t+\tau)u(t)\mathrm{d}t. \tag{1.3}$$

Power spectral density (PSD) of a stationary stochastic process is defined as:

$$\Phi_u(f) = \Im[\phi_u(\tau)] \tag{1.4}$$

where $\Im$ is the Fourier transform operator.

For an ergodic stochastic process, PSD is given by:

$$\Phi_u(f) = |U(f)|^2 \tag{1.5}$$

with

$$U(f) = \Im[u(t)]. \tag{1.6}$$

From Eq. 1.6 it is clear that PSD can be estimated by the Fourier transform of any realization if the stochastic process is ergodic and stationary.

Integrating a stochastic process $u(t)$, we obtain another stochastic process $v(t)$ whose PSD is given by

$$\Phi_v(f) = \frac{1}{(2\pi f)^2}\Phi_u(f). \tag{1.7}$$

When a white noise stochastic process is integrated, a random walk stochastic process (also called Winer–Lévy stochastic process) is obtained [17]. The main feature of the random walk stochastic process is that its variance is proportional to time. A random walk stochastic process models very well the measurement noise on estimation of the rotation angle.

In the PSD log–log plot of data obtained by measuring a constant angular rate (Fig. 1.1), we can distinguish three regions having different slope. The region

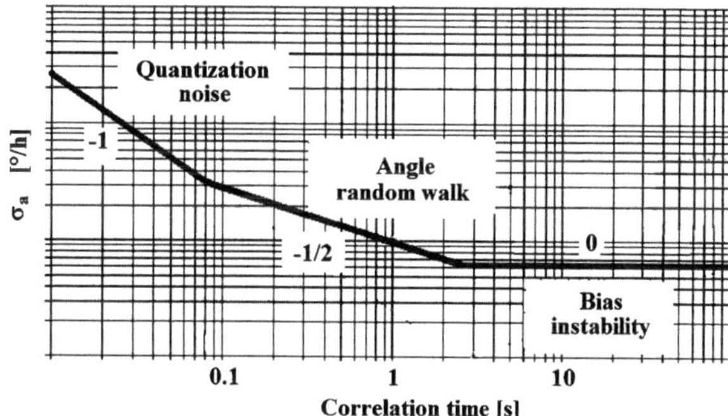

Fig. 1.2 Allen standard deviation of data obtained by measuring a constant angular rate (piecewise representation)

having slope equal to 0 is relevant to angle random walk. This means that the random walk noise is a white noise contribution to the angular velocity estimation. This can be explained observing that the rotation angle can be calculated by integrating the angular rate and, as previously explained, a random walk stochastic process is obtained by integrating a white noise stochastic process.

The region with -1 slope is relevant to bias instability and the region related to quantization noise has a slope equal to $+2$.

Gyroscope angle random walk coefficient W_{ARW} in $°/\sqrt{h}$ is given by [18, 19]:

$$W_{\mathrm{ARW}} = \frac{1}{60}\sqrt{\frac{A_{\mathrm{w}}}{2}} \tag{1.8}$$

where A_{w} is the white noise level of the gyroscope expressed in $(°/h)^2/\mathrm{Hz}$.

Alternatively the amount of noise contributions can be estimated by the Allan variance σ_{a}^2 [20] of data obtained by measuring a constant angular velocity [20, 22].

Let us denote with Ω_ξ ($h = 1,\ldots, N$) the angular rate data taken at a rate of f_{s} samples per second. From these N data we can form $K = N/M$ clusters (M is the number of samples per cluster). We can calculate the average for each cluster:

$$\bar{\Omega}_j(M) = \frac{1}{M}\sum_{\sigma=1}^{M}\Omega_{(j-1)M+\sigma} \quad (j = 1,\ldots, K) \tag{1.9}$$

and the Allan variance of angular rate data:

$$\sigma_{\mathrm{a}}^2(\tau_{\mathrm{a}}) = \frac{1}{2(K-1)}\sum_{j=1}^{K-1}\left[\bar{\Omega}_{j+1}(M) - \bar{\Omega}_j(M)\right]^2 \tag{1.10}$$

where $\tau_{\mathrm{a}} = M/f_{\mathrm{s}}$ is the correlation time.

In the log–log plot of $\sigma_{\mathrm{a}}(\tau)$ versus τ_{a} (see Fig. 1.2), the ARW can be estimated by the plot section having slope equal to $-1/2$, bias instability by the plot section having slope equal to 0 and quantization can be estimated by the plot section having slope equal to -1.

Table 1.1 Performance requirements for different classes of gyroscopes

Parameter	Rate grade	Tactical grade	Inertial grade
Angle random walk ($°/\sqrt{h}$)	>0.5	0.5–0.05	<0.001
Bias drift ($°/h$)	10–1000	0.1–10	<0.01
Scale factor accuracy (%)	0.1–1	0.01–0.1	<0.001
Full range ($°/h$)	1.5×10^5 to 3.6×10^6	$>1.8 \times 10^6$	$>1.4 \times 10^6$
Bandwidth (Hz)	>70	~ 100	~ 100

Based on the performance, we can distinguish three different classes of gyroscopes. So we classify inertial-grade, tactical-grade, and rate-grade gyros. Table 1.1 summarizes the gyro performance for each of these categories [23].

Obviously different applications require specific gyro performance. For instance, gyroscopes used in automotive applications requires a full range of at least 1.8×10^5 °/h (=50 °/s), a resolution of about 360 °/h (=0.1 °/s) and a bandwidth of 50 Hz, whereas other applications as autonomous navigation require better performance. Strategic missiles navigation requires a scale factor stability around 10 ppm and a bias stability around 1×10^{-4} °/h. Autonomous navigation of submarines requires a scale factor stability around 1 ppm and a bias stability around 1×10^{-3} °/h.

1.3 Gyro Applications

The fundamental application of gyros is within strapdown inertial navigation systems, used for navigation of ships, submarines, aircraft, guided missiles and other military vehicles. These navigation systems are directly mounted on the vehicle and allow its position and velocity estimation without the support of any signal generated by positioning systems, e.g. GPS (Global Positioning System).

Two basic building blocks of a strapdown inertial navigation system are the IMU and the navigation micro-computer. The IMU, including high performance gyros and accelerometers, measures vehicle angular rate and acceleration. The micro-computer processes data provided by the IMU and estimates vehicle position and velocity.

Satellites orientation is controlled by Attitude and Orbit Control Systems (AOCSs) including different attitude sensors such as gyros, sun sensors, earth sensors, star trackers, magnetometers and so on. Main space application of angular rate sensors is within AOCSs but recently gyros have been successfully exploited for navigation of rover vehicles, too. In particular, rovers developed by NASA and Jet Propulsion Laboratory for Mars exploration were equipped with FOGs [24]. Space applications require gyros having a resolution in the range 0.01–10 °/h.

Antijitter platforms in high quality digital cameras, GPS backup systems, virtual reality devices and gaming consoles are some typical examples of consumer electronics products using low cost gyros having quite limited performance. Usually gyros for consumer electronics are MEMS devices.

The widest market for low cost angular rate sensors is automotive. Traction control systems, active suspensions and anti-skidding systems include gyros typically realized by silicon MEMS technologies.

Finally, robotics and medicine are currently emerging as new gyro applications.

References

1. Murugesan, S., Goel, P.S.: Autonomous fault-tolerant attitude reference system using DTGs in symmetrically skewed configuration. IEEE Trans. Aerosp. Electron. Syst. **25**, 302–307 (1989)
2. Lappas, V.J., Steyn, W.H., Underwood, C.I.: Torque amplification of control moment gyros. Electron. Lett. **38**, 837–839 (2002)
3. Defendini, A., Faucheux, P., Guay, P., Morand, J., Heimel, H.: A compact CMG product for agile satellite. In: 5th ESA Conference on Spacecraft Guidance, Navigation and Control, Frascati (Rome), Italy, 22–25 October 2002
4. Izmailov, E.A., Kolesnik, M.M., Osipov, A.M., Akimov, A.V.: Hemispherical resonator gyro technology. Problems and possible ways of their solutions. In: RTO SCI International Conference on Integrated Navigation Systems, St. Petersburg, Russia, 24–26 May 1999
5. Jourdan, D.: MEMS gyroscope market is expected to reach 800 M$ in 2010. Sens. Transducers **67** (2006)
6. Macek, W.M., Davis, D.T.M.: Rotation rate sensing with traveling-wave ring lasers. Appl. Phys. Lett. **2**, 67–68 (1963)
7. Ciminelli, C.: Innovative photonic technologies for gyroscope systems. In: EOS Topical Meeting—Photonic Devices in Space, Paris, France, 16–19 October 2006
8. Ciminelli, C., Peluso, F., Armenise, M.N.: A new integrated optical angular velocity sensor. Proc. SPIE **5728**, 93–100 (2005)
9. Unger, G.L., Kaufman, D.M., Krainak, M.A., Sanders, G.A., Taylor, W.L., Schulze, N.R.: NASA's first in-space optical gyroscope: a technology experiment on the X-ray Timing Explorer spacecraft. Proc. SPIE **1953**, 52–58 (1993)
10. Woodman, K.F., Franks, P.W., Richards, M.D.: The nuclear magnetic resonance gyroscope—a review. J. Navig. **40**, 366–384 (1987)
11. Bruckner, N., Packard, R.: Large area multiturn superfluid phase slip gyroscope. J. Appl. Phys. **93**, 1798–1805 (2003)
12. Titterton, D.H., Weston, J.L.: Strapdown Inertial Navigation Technology. IET (2005)
13. Barbour, N.: Inertial navigation sensors. In: Advances in Navigation Sensors and Integration Technology. NATO RTO Educational Notes (2004)
14. Armenise, M.N., Ciminelli, C., De Leonardis, F., Diana, R., Passaro, V., Peluso, F.: Gyroscope technologies for space applications. In: 4th Round Table on Micro/Nano Technologies for Space, Noordwijk, The Netherlands, 20–22 May 2003
15. European Space Agency (ESA), IOLG project 1678/02/NL/PA: Micro gyroscope technologies for space applications. Contract Report, June 2003
16. IEEE Standard for Inertial Sensor Terminology (Std 528-2001)
17. Papoulis, A., Pillai, S.U.: Probability, Random Variables and Stochastic Processes. McGraw-Hill, New York (2001)
18. IEEE Recommended Practice for Inertial Sensor Test Equipment, Instrumentation, Data Acquisition, and Analysis (Std 1554-2005)
19. IEEE Standard Specification Format Guide and Test Procedure for Linear, Single-Axis, Non gyroscopic Accelerometers (Std 1293-1998)
20. Percival, D.B., Walden, A.T.: Wavelet Methods for Time Series Analysis. Cambridge University Press, Cambridge (2006)

21. IEEE Standard Specification Format Guide and Test Procedure for Single-Axis Interferometric Fiber OpticGyros (Std 952-1997)
22. Ng, L.C., Pines, D.J.: Characterization of ring laser gyro performance using the Allan variance method. J. Guid. Control Dyn. **20**, 211–214 (1996)
23. Lawrence, A.: Modern Inertial Technology: Navigation, Guidance, and Control. Springer, New York (1998)
24. Ali, K.S., Vanelli, C.A., Biesiadecki, J.J., Maimone, M.W., Cheng, Y., San Martin, A.M., Alexander, J.W.: Attitude and position estimation on the mars exploration rovers. In: IEEE Systems, Man and Cybernetics, International Conference, Waikoloa, Hawaii, USA, 10–12 October 2005

Chapter 2
Physical Effects in Gyroscopes

2.1 Sagnac Effect

Operating principle of all optical gyros is based on Sagnac effect [1]. It induces either a phase shift $\Delta\varphi$ between two optical signals propagating in opposite directions within a ring interferometer rotating around an axis perpendicular to the ring, or a frequency shift between two resonant modes propagating in clockwise (CW) and counter-clockwise (CCW) directions within an optical cavity rotating around an axis perpendicular to it.

To derive the analytic expression of the rotation induced phase shift between CW and CCW beams, a simple cinematic approach can be used [2]. We first consider a circular ring interferometer in which two waves counter-propagate in the vacuum (see Fig. 2.1). Light enters the interferometer at point P and is split into CW and CCW propagating signals by a beam splitter. When the interferometer is at rest with respect to a motionless inertial frame of reference, optical path lengths of the two optical signals propagating in opposite directions (CW and CCW signals) are equal. Also the speeds of the two signals are equal to c (c is light speed in the free space). After propagating in the loop both waves come back into the beam splitter after a time interval τ_r equal to:

$$\tau_r = \frac{2\pi R}{c} \tag{2.1}$$

where R is the ring interferometer radius.

If the ring interferometer is rotating at a rate Ω, the beam splitter located in P moves during the time interval τ_r by a length $\Delta l = \Omega R \tau_r$.

CW (co-directional with Ω) beam experiences a path length slightly greater than $2\pi R$ in order to complete one round trip, since the ring interferometer rotates through a small angle during the round-trip transit time. CCW beam experiences a path length slightly less than $2\pi R$ during one round trip. The difference between optical paths of CW (L_{CW}) and CCW (L_{CCW}) waves is given by.

M. N. Armenise et al., *Advances in Gyroscope Technologies*,
DOI: 10.1007/978-3-642-15494-2_2, © Springer-Verlag Berlin Heidelberg 2010

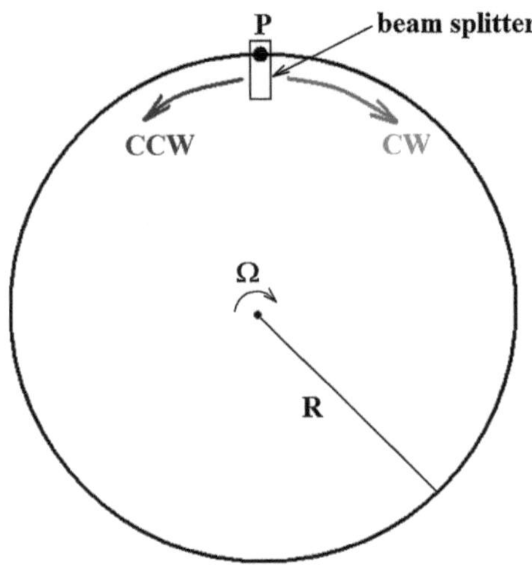

Fig. 2.1 Sagnac ring interferometer

$$\Delta L = L_{CW} - L_{CCW} = 2\Delta l = 2\Omega R \tau_r = \frac{4\pi\Omega R^2}{c} \tag{2.2}$$

Since CW and CCW waves propagate at the same speed (equal to the speed of light in vacuum c), the CCW wave arrives in P before the CW wave. The delay between the two optical signals is equal to:

$$\Delta t = \frac{\Delta L}{c} = \frac{4\pi\Omega R^2}{c^2} \tag{2.3}$$

The phase shift $\Delta\varphi$ between CW and CCW optical signals due to the interferometer rotation can be written as:

$$\Delta\varphi = \Delta t \frac{2\pi c}{\lambda} = \frac{8\pi^2 R^2}{c\lambda}\Omega \tag{2.4}$$

where λ is the optical signal wavelength.

Phase shift expression in Eq. 2.4 is valid for one circular loop. If optical path of CW and CCW beams consists of k turns, the phase shift $\Delta\varphi$ becomes:

$$\Delta\varphi = \frac{8\pi^2 R^2}{c\lambda}k\Omega \tag{2.5}$$

The time delay expression reported in Eq. 2.3 can be also derived in the framework of special relativity [3].

Now we consider a similar interferometer in which the vacuum is substituted by a homogeneous dielectric medium having a refractive index equal to n. While the interferometer is at rest, light travels at the speed c/n in both directions and propagation time around the loop of both waves is equal to $n \cdot \tau_r$. Both waves are still in phase after a propagation time equal to $n \cdot \tau_r$.

If the circular path is rotating, the beam splitter located in P has moved through a length $n\Delta l$ in the propagation time $n \cdot \tau_r$. So the optical path length of CW wave in one round trip is equal to:

$$L_{\text{CW}}^* = 2\pi R + n\Delta l = 2\pi R + \frac{2\pi n \Omega R^2}{c} \tag{2.6}$$

whereas the optical path length of CCW wave in one round trip is given by:

$$L_{\text{CCW}}^* = 2\pi R - n\Delta l = 2\pi R - \frac{2\pi n \Omega R^2}{c} \tag{2.7}$$

In this case the speed of light is no longer the same for both counter-propagating signals. In particular the speed of CW wave is equal to:

$$\upsilon_{\text{CW}} = \frac{c}{n} + \alpha_d \Omega R \tag{2.8}$$

and the speed of CCW wave is:

$$\upsilon_{\text{CCW}} = \frac{c}{n} - \alpha_d \Omega R \tag{2.9}$$

where α_d is the Fresnel–Fizeau drag coefficient, which is given by [4]:

$$\alpha_d = 1 - n^{-2} \tag{2.10}$$

The additive terms in light velocity expressions in Eqs. 2.8 and 2.9 are due to the drag of light propagating in a uniformly moving medium [5].

The CW and CCW waves arrive in P in different instants. The delay between these two time instants is equal to:

$$\Delta t^* = \frac{L_{\text{CW}}^*}{\upsilon_{\text{CW}}} - \frac{L_{\text{CCW}}^*}{\upsilon_{\text{CCW}}} = \frac{2\pi R + \frac{2\pi n \Omega R^2}{c}}{\frac{c}{n} + \alpha_d \Omega R} - \frac{2\pi R - \frac{2\pi n \Omega R^2}{c}}{\frac{c}{n} - \alpha_d \Omega R} \tag{2.11}$$

Rearranging Eq. 2.11 and assuming $c^2/n^2 \gg \alpha_d \Omega^2 R^2$ we obtain:

$$\Delta t^* \cong \frac{4\pi R^2 n^2 \Omega (1 - \alpha_d)}{c^2} = \frac{4\pi R^2 \Omega}{c^2} \tag{2.12}$$

By comparing Eqs. 2.12 and 2.3, one can conclude that $\Delta t = \Delta t^*$. Therefore, the phase shift induced by rotation is equal when optical propagation takes place in the vacuum or in a homogeneous medium having refractive index equal to n. The same result can be demonstrated using a more rigorous relativistic electrodynamic approach which consists of the derivation of the equation describing the optical propagation in a rotating frame and the application of a method of perturbation to calculate the rotation induced phase shift [6].

As previously pointed out, rotation induces a frequency difference between two counter-propagating resonant modes excited in an optical cavity

An optical resonator at rest, in which optical propagation occurs in the vacuum, supports optical modes whose resonance frequencies $v_{q,0}$ satisfy the following relation:

$$qc = v_{q,0}\,p \tag{2.13}$$

where q is an integer number (resonance order) and p is the perimeter of the resonator. If two qth order counter-propagating resonant modes are excited in a ring cavity, their resonance frequencies are split by rotation being equal to:

$$v_q^{\text{CW}} = \frac{qc}{p_+} \qquad v_q^{\text{CCW}} = \frac{qc}{p_-} \tag{2.14}$$

where p_+ and p_- are perimeters of optical paths experienced by the two resonant modes. The difference $p_+ - p_-$ is denoted with Δp.

The frequency difference Δv between the two qth order resonant modes is given by:

$$\Delta v = v_q^{\text{CCW}} - v_q^{\text{CW}} = qc\left(\frac{1}{p_-} - \frac{1}{p_+}\right) \cong qc\frac{\Delta p}{p^2} \tag{2.15}$$

Combining Eqs. 2.15 and 2.13, we obtain:

$$\Delta v = v_{q,0}\frac{\Delta p}{p} \tag{2.16}$$

The expression of the frequency splitting given by Eq. 2.16 does not change if optical propagation within the resonator occurs in a homogeneous medium having refractive index n or an optical waveguide having effective index n_{eff}.

For a circular ring resonator $\Delta p = 4\pi\Omega R^2/c$, p is obviously equal to $2\pi R$ and so Δv can be written as:

$$\Delta v = v_{q,0}\frac{2R}{c}\Omega \tag{2.17}$$

where R is the resonator radius.

If the optical cavity has an arbitrary geometry, the frequency difference Δv results [7]:

$$\Delta v = \frac{4a v_{q,0}}{pc}\Omega \tag{2.18}$$

where a is the area enclosed by the light path.

2.2 Coriolis Force Effect

The operating principle of all vibrating gyros is based on the effect of Coriolis force on a vibrating mass. A simple model of vibrating angular rate sensors is a two degree-of-freedom spring-mass-damper system shown in Fig. 2.2.

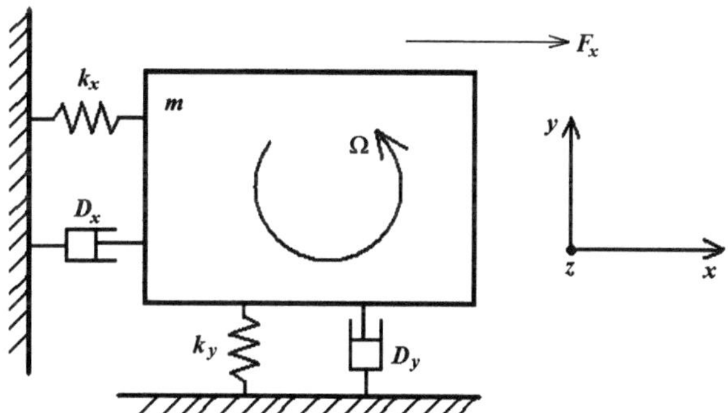

Fig. 2.2 Two degree-of-freedom spring-mass-damper system in a rotating reference frame

Coriolis force is a fictitious force experienced by a mass m moving in a rotating reference frame. It is equal to:

$$\overrightarrow{F_c} = 2m\left(\overrightarrow{v} \times \overrightarrow{\Omega}\right) \tag{2.19}$$

where $\overrightarrow{v}$ is the mass velocity in the rotating reference frame and $\overrightarrow{\Omega}$ is the angular velocity of the reference frame. The effect of Coriolis force on the two degree-of-freedom spring-mass-damper system shown in Fig. 2.2 can be derived starting from dynamic equations describing the motion of the system in a rotating reference frame.

In Fig. 2.2, the mass m can move along x and y axes and $\overrightarrow{\Omega}$ is directed along z. The oscillation along x, namely drive or primary oscillating mode, is excited by the force F_x directed along x whereas the oscillation along y, sense or secondary oscillating mode, is due to system rotation around z axis. Motion equations of the two degree-of-freedom system can be written in the form [8]:

$$\begin{cases} m\frac{d^2x}{dt^2} + D_x\frac{dx}{dt} + k_x x - 2\Omega m\frac{dy}{dt} = F_x \\ m\frac{d^2y}{dt^2} + D_y\frac{dy}{dt} + k_y y + 2\Omega m\frac{dx}{dt} = 0 \end{cases} \tag{2.20}$$

where Ω is the module of reference system angular rate, D_x and D_y are the damping coefficients along x and y axes, and k_x and k_y are the spring constants along x and y axes.

Usually the primary oscillating mode is excited by a sinusoidal force F_x and its amplitude is kept constant at a_x. To maximize a_x, the angular frequency of the exciting force ω_d is typically very close to the resonance frequency $\omega_x = \sqrt{k_x/m}$ of the primary resonator. So $x(t)$ can be written as:

$$x(t) = a_x \sin(\omega_d t) \cong a_x \sin(\omega_x t) \tag{2.21}$$

To calculate $y(t)$ we can use the second equation of the system reported in Eq. 2.20. This equation can be rewritten in the form:

$$\frac{d^2y}{dt^2} + \frac{\omega_y}{Q_y}\frac{d^2y}{dt^2} + \omega_y^2 y = -2a_x\Omega\omega_x\cos(\omega_x t) \tag{2.22}$$

where $\omega_y = \sqrt{k_y/m}$ is the resonance frequency of the secondary resonator and $Q_y = \sqrt{mk_y}/D_y$ is the quality factor of the sense mode. After the transient regime, $y(t)$ assumes the general form:

$$y(t) = a_y\cos\left[\omega_x t + \phi_y\right] \tag{2.23}$$

where a_y and ϕ_y are the amplitude and the phase response of the secondary resonator at ω_x, respectively.

Calculating dy/dt and d^2y/dt^2 and substituting them in Eq. 2.22, we obtain:

$$\left[-a_y\omega_x^2\cos(\phi_y) + a_y\omega_y^2\cos(\phi_y) - \frac{a_y\omega_x\omega_y}{Q_y}\sin(\phi_y)\right]\cos(\omega_x t)$$
$$+ \left[a_y\omega_x^2\sin(\phi_y) - a_y\omega_y^2\sin(\phi_y) - \frac{a_y\omega_x\omega_y}{Q_y}\cos(\phi_y)\right]\sin(\omega_x t)$$
$$= -2a_x\Omega\omega_x\cos(\omega_x t) \tag{2.24}$$

Equation 2.24 provides the following algebraic system:

$$\begin{cases} \left[-a_y\omega_x^2\cos(\phi_y) + a_y\omega_y^2\cos(\phi_y) - \frac{a_y\omega_x\omega_y}{Q_y}\sin(\phi_y)\right] = -2a_x\Omega\omega_x \\ \left[a_y\omega_x^2\sin(\phi_y) - a_y\omega_y^2\sin(\phi_y) - \frac{a_y\omega_x\omega_y}{Q_y}\cos(\phi_y)\right] = 0 \end{cases} \tag{2.25}$$

By solving the equation system (2.25), a_y and $y(t)$ are derived as follows:

$$a_y = -\frac{2a_x\Omega\omega_x}{\sqrt{\left(\omega_x^2 - \omega_y^2\right)^2 + \omega_x^2\omega_y^2/Q_y^2}} \tag{2.26}$$

$$y(t) = -\frac{2a_x\Omega\omega_x}{\sqrt{\left(\omega_x^2 - \omega_y^2\right)^2 + \omega_x^2\omega_y^2/Q_y^2}}\cos(\omega_x t + \phi_y) \tag{2.27}$$

Equation 2.27 shows that the amplitude of sense mode is directly proportional to the angular rate Ω. Then the angular rate of the two degree-of-freedom spring-mass-damper system can be easily estimated by measuring the amplitude of the oscillation along y.

References

1. Sagnac, G.: L'èther lumineux dèmontrè par l'effet du vent relatif d'èther dans un interfèromètre en rotation uniforme. C. R. Acad. Sci. **95**, 708–710 (1913)
2. Arditty, H.J., Lefevre, H.C.: Sagnac effect in fiber gyroscopes. Opt. Lett. **6**, 401–403.
3. Rizzi, G., Ruggiero, M.L.: A direct kinematical derivation of the relativistic Sagnac effect for light or matter beams. Gen. Relativ. Gravit. **35**, 2129–2136 (2003)
4. Vali, V., Shorthill, R.W., Berg, M.F.: Fresnel–Fizeau effect in a rotating optical fiber ring interferometer. Appl. Opt. **16**, 2605–2607 (1977)
5. Drezet, A.: The physical origin of the Fresnel drag of light by a moving dielectric medium. Eur. Phys. J. **B45**, 103–110 (2005)
6. Lefevre, H.C., Arditty, H.J.: Electromagnetisme des milieux dielectriques lineaires en rotation et application a la propagation d'ondes guidees. Appl. Opt. **21**, 1400–1409 (1982)
7. Jacobs, F., Zamoni, R.: Laser ring gyro of arbitrary shape and rotation axis. Am. J. Phys. **50**, 659–660 (1982)
8. Acar, C.: Robust micromachined vibratory gyroscopes. PhD dissertation, University of California, Irvine, California, USA (2004)

Chapter 3
He–Ne and Solid-State Ring Laser Gyroscopes

Commercial success of the He–Ne Ring Laser Gyroscope (RLG) began in the late 1980s and early 1990s. Since He–Ne RLG first demonstration in 1963 [1], a number of industrial companies have developed a great research effort for RLG technology improvement so that this optical sensor has become a widely diffused commercial device. For instance, navigation systems based on He–Ne RLGs has been installed on over 50 different aircrafts [2]. Since several years, He–Ne RLG dominates the high-performance gyros market.

Main advantages of the He–Ne RLG with respect to mechanical gyros are the absence of moving parts, the design simplicity (less than 20 components), the insensitivity to vibrations, the digital output, the wide dynamic range, the fast update rate, and the good reliability. In the last years, the replacement of the He–Ne gain medium with a solid-state one has been proposed by some researchers who swear that a RLG having a solid-state gain medium could exhibit longer lifetime, lower cost and simpler manufacturing process with respect to the He–Ne RLG [3]. This innovative optical gyroscope has been called solid-state RLG. As the He–Ne RLG, the solid-state RLG includes a number of discrete optical components and so it is not an integrated optical device.

The RLG is a bulk-optics sensor and its operating principle, valid for either He–Ne or solid-state RLG, was proposed by Rosenthal [4] for the first time (see Fig. 3.1). A gain medium, producing optical amplification of light passing through it, is placed within an optical resonant cavity. Two counter-propagating cavity resonant modes are excited within the cavity and difference between their frequencies is observed. When the whole system rotates with angular rate equal to Ω, this frequency difference is proportional to Ω.

A wide review of He–Ne RLG operating principle and the research effort developed during the 1960s to enhance performance of this kind of active optical gyro is reported in [5]. More recent reviews on He–Ne RLGs are available in [6–8].

M. N. Armenise et al., *Advances in Gyroscope Technologies*,
DOI: 10.1007/978-3-642-15494-2_3, © Springer-Verlag Berlin Heidelberg 2010

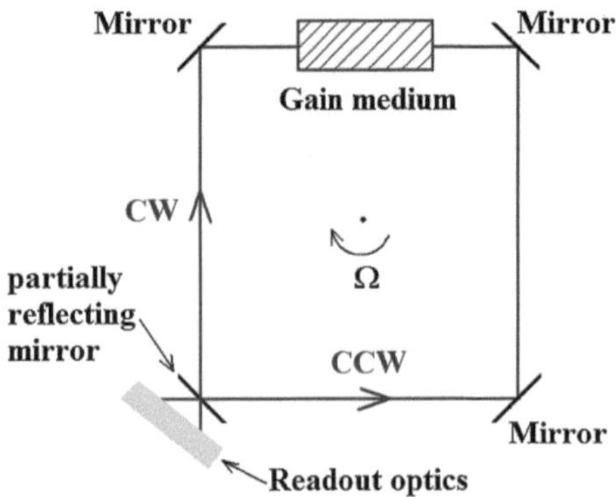

Fig. 3.1 RLG general architecture

In this chapter, He–Ne RLG architectures are illustrated and noise and error sources are described. Finally operating principle and architecture of the bulk-optic solid-state RLG are also discussed.

3.1 He–Ne RLG Architectures

The He–Ne RLG has been both theoretically and experimentally widely investigated. Two main architectures for its realization have been proposed. The fundamental difference between them is the shape of the optical cavity in which the counter-propagating laser beams are excited. In the first architecture, shown in Fig. 3.2a, two corner mirrors and a spherical mirror have been used to realize an equilateral triangular optical cavity [9]. In the second architecture (see Fig. 3.2b), exploited in the first proposed He–Ne RLG and in some commercially available RLGs, a square optical cavity realized by four corner mirrors has been used. In 1987, Lim et al. [10] patented a five sided He–Ne RLG.

To monitor the angular rate around each axis of a Cartesian reference system, three gyros are mounted on a vehicle. A configuration with three He–Ne RLGs in a single cubic block has been patented [11]. The architecture of this sensor includes three mutually orthogonal squared He–Ne RLGs which share two mirrors so that only six mirrors are needed. Also the cathode and the dithering system is shared among the three He–Ne RLGs.

To increase He–Ne RLG mechanical stability and decrease the thermal sensitivity, the whole sensor is usually realized within a solid block of a material with a very low coefficient of thermal expansion (for example a vitro-ceramic glass).

The gain medium is a gas mixture in which He pressure is around 10 mbar and the He:Ne mixing ratio is about 10:1. This gas mixture is contained inside a discharge tube equipped with electrodes (anode and cathode) to supply electrical power. When the laser is turned on, an initial voltage pulse of 7–8 kV is applied across the electrodes to ionize the gas. During lasing, a voltage of 1–2 kV is applied across the electrodes producing a current of several mA through the laser

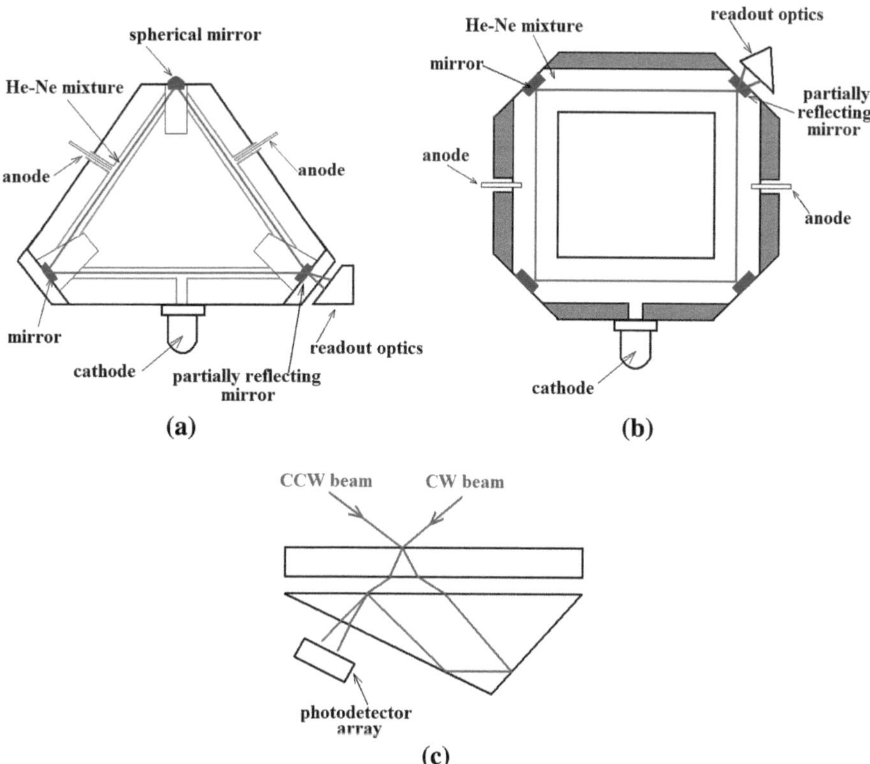

Fig. 3.2 **a** He–Ne RLG architecture based on a triangular optical cavity. **b** He–Ne RLG architecture based on a squared optical cavity. **c** Readout optics exploited in He–Ne RLGs

tube. Electrons within the gas are accelerated and excite He atoms, which populate metastable He energy levels. Some of this energy is transferred, by atom collision, to Ne atoms, which populate energy levels having excitation energy similar to that of metastable He energy levels. Excited Ne atoms radiatively decay to lower Ne energy states generating CW and CCW optical signals. To assure that the two resonant modes coexist in the active cavity and have the same optical power level, the mode competition between them has to be avoided.

In the He–Ne ring laser, mode competition between the oppositely directed waves has been observed when only one Ne isotope is present in the gas mixture. Using a natural neon mixture (91% ^{20}Ne : 9% ^{22}Ne) in the discharge tube mode competition is avoided if operating wavelength is equal to 1.15 µm. For the He–Ne RLG operating at 633 nm, mode competition can still occur when ^{22}Ne percentage in neon mixture is around 10%. A further increase of ^{22}Ne percentage allows to avoid mode competition also when operating frequency is 633 nm [12, 13]. In fact, in commercial He–Ne RLGs, the neon is a 50:50 mixture of ^{22}Ne and ^{20}Ne isotopes. The degradation of He–Ne gas mixture quality can affect the sensor lifetime so, in RGL design, it is necessary to limit the number of openings within the discharge tube through which the gas can leak.

The He–Ne RLG rotation induces a frequency difference between CW and CCW laser beams. The most used optical readout apparatus for the He–Ne RLG includes a combining prism and a photodetection apparatus (Fig. 3.2c). The prism allows to combine CW and CCW optical signals. The beams coming out from the prism are quite collinear and interfere forming a fringe pattern. The time-dependent intensity of light on the surface of the detector array is given by:

$$I(x, t) = I_0\left[1 + \cos\left(2\pi\Delta v t + \frac{2\pi}{\lambda}\gamma_p x + \varphi_0\right)\right] \tag{3.1}$$

where I_0 is the mean value of $I(x, t)$, γ_p is the angle between the beams at the output of the prism, φ_0 is a constant phase difference between the beams, λ is the laser operating wavelength, x is the spatial coordinate measured along the detector array, and Δv is the rotation-induced frequency difference between CW and CCW beams. When the RLG rotates, the fringe pattern moves in a direction depending on the sense of rotation.

For a given integration time $\Delta\tau_i$, the number N_m of intensity maxima counted by the detector array is:

$$N_m = \int_0^{\Delta\tau_i} \Delta v dt \tag{3.2}$$

Since $\Delta v = (4\,a/p\lambda)\,\Omega$, we can write N_m as:

$$N_m = \frac{4a}{p\lambda}\int_0^{\Delta\tau_i} \Omega dt = \frac{4a}{p\lambda}\theta = S\theta \tag{3.3}$$

where a and p are the area and the perimeter of the optical cavity (respectively), θ is the gyro total rotation angle during the time interval $\Delta\tau_i$ and $S = 4\,a/p\lambda$ is the scale factor of all gyros based on an active or passive optical cavity. Equation 3.3 shows the linear relationship between the total rotation angle and the number of fringes counted by the detector.

3.2 Error Sources in He–Ne RLGs

Ideally, in the He–Ne RLG, the relationship between angular rate Ω and frequency shift Δv is perfectly linear. Any effect inducing a deviation from this linear relation can be considered as a source of error, which may limit the RLG performance. In He–Ne RLGs, three main error sources are usually considered because of their relevance: null shift (or bias), mode locking and scale factor variations. A review of error sources in He–Ne RLGs has been reported in [14].

3.2.1 Null Shift

Null shift is a nonzero frequency difference between CW and CCW beams for a zero input angular rate. Taking into account this error source, the relation between angular rate Ω and frequency shift Δv can be written as:

$$\Delta v = \frac{4a}{p\lambda}\Omega + K_0 = S\Omega + K_0 \qquad (3.4)$$

where K_0 is the bias term. Exact magnitude of the bias term is usually unpredictable and time varying. Bias drift in high-quality He–Ne RLG is around 0.01 °/h.

Bias drift depends on localized anisotropy sources in the cavity with respect to optical signals traveling in the two opposite directions. The main source of anisotropy is the so-called Langmuir flow in the active laser medium [15]. In a DC-excited plasma, there is a movement of the neutral atoms, which is towards the cathode along the center of the discharge tube and towards the anode along the walls of the same tube (see Fig. 3.3). The laser energy is concentrated in the center of the tube and therefore it passes through the gas mixture flowing towards the cathode. The flow causes a shift in the refraction index that depends on the relative directions of the laser energy and the gas flow. This shift induces an anisotropy in the resonant cavity that can be significantly reduced by using a balanced configuration with two anodes and one cathode. This balanced configuration is used in all commercially available He–Ne RLGs.

3.2.2 Lock-in

Mode locking (or lock-in) is a detrimental effect leading to the vanishing of frequency difference between counter-propagating beams for small rotation rates. When the angular rate is below a critical value, mutual coupling of the two waves propagating in opposite directions locks them together. Then counter-propagating

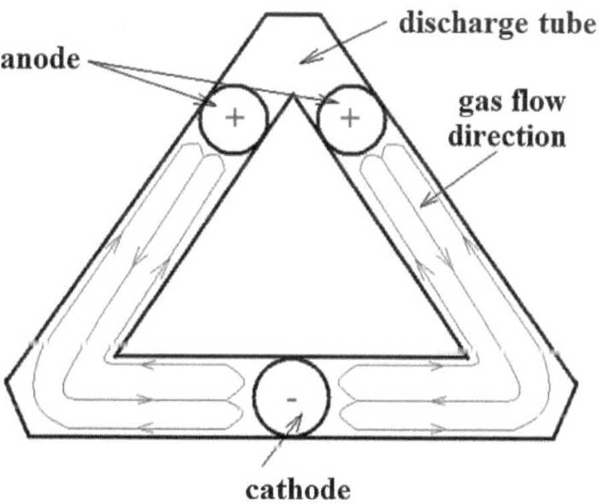

Fig. 3.3 Gas flow in a He–Ne RLG having a balanced configuration with two anodes and one cathode

beams oscillate at the same frequency, making $\Delta v = 0$. The coupling between the counter-propagating waves is mainly due to the fact that a small fraction of optical power carried by both the waves is backscattered. The physical origin of back-scattering is related to imperfections in mirrors or other intra-cavity optical elements and so, to reduce the influence of this physical effect, fabrication technology of mirrors has to be particularly accurate. The high accuracy required for mirror realization and polish tends to make the He–Ne RLG quite expensive. In 1970s and 1980s a significant research effort has been developed to theoretically describe mode locking mechanism [16–19].

According with semiclassical laser theory [20], in presence of backscattering, the total phase shift ψ between CW and CCW optical signals can be modeled by the following differential equation [21]:

$$\frac{d\psi}{dt} = S\Omega + b[sin(\psi)] \tag{3.5}$$

where b is the backscattering coefficient which has units of frequency and takes into account all back-reflection within the active cavity. If $b << S\Omega$ (large Ω values), no stationary solution of Eq. 3.5 exists. In this case frequency difference between CW and CCW signals linearly depends on rotation rate. If b is comparable to $S\Omega$ (small Ω values), the differential equation (3.5) exhibits a stationary solution, and the frequency difference between CW and CCW signals vanishes. Under these conditions a dead band extending from $-\Omega_L$ to $+\Omega_L$ (Ω_L is the dead band delimiting rate) arises within the RLG static characteristic shown in Fig. 3.4.

To reduce detrimental effects of the lock-in, an externally controlled constant bias to let the RLG always operating in the unlocked region can be added to the actual angular velocity. A number of techniques has been proposed to produce this constant bias such as the physical rotation of the gyroscope [22], the insertion of passive elements to generate a constant difference between optical paths of counter-propagating waves [23], the introduction of a Faraday cell within the ring cavity [24, 25] and the use of magnetic mirrors [26–28]. For typical Ω_L values, the

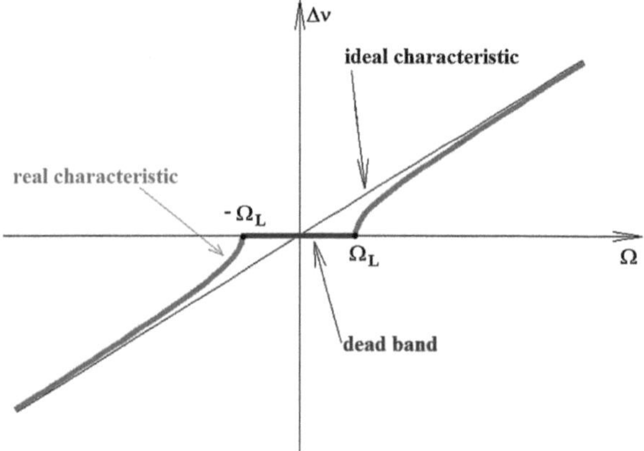

Fig. 3.4 Mode locking effect on He–Ne RLG static characteristic

constant bias would be of about 10^6 °/h but it is very difficult to keep constant so large bias value.

This problem can be solved by imposing a time-varying bias driven by a sine-wave signal. The most used technique to generate this bias is the mechanical one. In this case the gyro is rotated alternately in one direction and in the opposite. This approach, called mechanical dithering, is usually implemented by mounting the RLG on a rotating system which oscillates by means of a piezoelectric transducer. An alternative technique to impose this oscillating bias is the use of a Faraday cell generating an alternating magnetic field.

The introduction of a time-varying alternating bias induces the so called dynamic lock-in, which implies that, for certain ranges of input rates, the RLG output follows the dither input rate and is not responsive to the angular rate. To overcome this problem, the random variation of the initial phase relevant to every dither bias cycle has been proposed [29]. This solution can be implemented by adding an amplitude noise to pure sine dithering signal. The introduction of this randomized sinusoidal dithering generates a contribution to RLG angle random walk ($W_{ARW,d}$) [30]:

$$W_{ARW,d} = \Omega_L \sqrt{\frac{S}{2\pi\Omega_D}} \tag{3.6}$$

where Ω_D is the amplitude of sinusoidal dithering signal. Equation 3.6 shows that, in presence of mechanical dithering, the ARW is proportional to the width of dead band due to lock-in (=2 Ω_L).

Lock-in issue can be significantly alleviated by using a ring cavity that supports more than one pair of counter-propagating modes [31]. The exploitation of a four resonant modes RLG, called multioscillator ring laser gyroscope, has been firstly proposed in [32]. In this gyro, whose architecture is shown in Fig. 3.5a, two counter-propagating right-circularly polarized (RCP) modes and two counter-propagating left-circularly polarized (LCP) modes propagate. A reciprocal bias is used to split the frequency of co-propagating modes having different circular polarizations whereas a non-reciprocal bias (equal to Δv_0) splits the resonant frequencies of counter-travelling modes having the same polarization (Fig. 3.5b). The frequency differences between LCP and RCP resonant modes are given by:

$$\Delta v_{LCP} = \Delta v_0 + S\Omega \tag{3.7}$$

$$\Delta v_{RCP} = \Delta v_0 - S\Omega \tag{3.8}$$

and the difference between Δv_{LCP} and Δv_{RCP} is equal to $2S\Omega$. In this sensor, the angular velocity is estimated by measuring $\Delta v_{LCP} - \Delta v_{RCP}$ and so the scale factor is equal to 2 S.

The non-reciprocal bias can be achieved by Faraday effect in a cell located inside the cavity (as in the Differential Laser Gyro [33] and in the Zero-Lock Laser Gyro [34]) or by Zeeman effect [35, 36] in the gain medium itself.

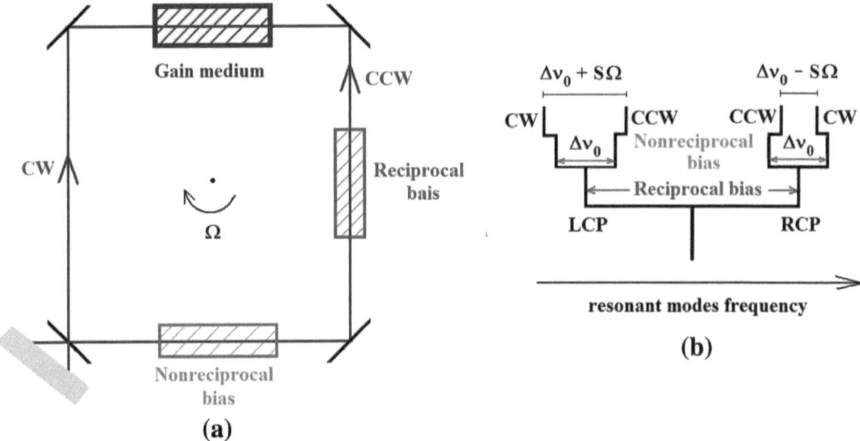

Fig. 3.5 **a** Basic architecture of the four-wave RLG. **b** Frequency diagram of the four-wave RLG

To reduce the mode locking, short light pulses excitation within the RLG has been proposed in [37, 38]. This approach increases the complexity of the RLG and does not allow a significant enhancement of sensor performance.

3.2.3 Scale Factor Variation

He–Ne RLG scale factor is not perfectly equal to the nominal one (S) given by ($4\ a/p\lambda$). According with [39], the experimentally measured gyro scale factor S^* can be expressed as:

$$S^* = S(1 - \delta) \tag{3.9}$$

where δ is a correction factor larger then zero which depends on ring laser output power, laser gain and gain medium dispersion. In He–Ne RLGs, deviation of the scale value from the nominal one is less than a few ppm.

3.3 Quantum Noise in He–Ne RLGs

Because of spontaneous emission, optical signals generated within a He–Ne RLG have a random phase, which can be modeled by a stochastic process. This random phase variation is the most important noise contribution observable at the gyro output. This noise contribution defines the maximum achievable resolution (quantum limit). Theoretical investigation of He–Ne RLG quantum limit has been developed in [40, 41]. In a typical He–Ne RLG operating at 633 nm and having a perimeter of 20 cm, the contribution to ARW due to quantum noise is approximately $2 \times 10^{-4}\,°/\sqrt{h}$.

In randomized sinusoidal dithered He–Ne RLG, the contribution to angle random walk coefficient due to dithering is about one order of magnitude larger than the contribution due to spontaneous emission ($ARW_d \approx 2 \times 10^{-3}\,°/\sqrt{h}$). In a He–Ne RLG without mechanical dithering, the angle random walk is only due to quantum noise. A RLG operating at quantum limit has been demonstrated in [42].

To improve the He–Ne RLG performance in terms of quantum noise and backscattering, the use of carbon dioxide as gain medium has been proposed in [43]. In fact CO_2 RLG can operate at higher power and longer wavelength (10.4 or 9.4 µm) with respect to He–Ne RLG. Because quantum limit decreases when intra-cavity power increases and also backscattering is reduced by increasing operating wavelength, CO_2 RLG could be a valid alternative to He–Ne RLG. Experimental data are not sufficient to compare CO_2 RLG performance with that of He–Ne RLG.

3.4 Solid-State RLGs

The gaseous gain medium exploited in the He–Ne RLG limits its reliability and lifetime. The replacement of the He–Ne gain medium with a solid-state one consisting of a Nd:YAG optical amplifier has been explored in [44]. In the

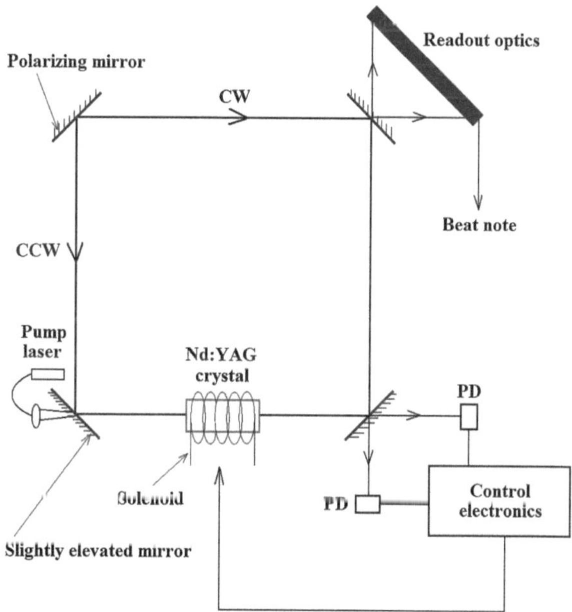

Fig. 3.6 Solid-state RLG architecture (PD: photodetector)

proposed architecture (shown in Fig. 3.6), the gain medium is inserted in a squared optical cavity formed by four mirrors. A bulk-optic solid-state ring laser operating at 1,064 nm is obtained. The laser is optically pumped by a laser diode at 808 nm. Generating two counter-propagating beams within this ring laser, the device may be used as gyroscope [45].

Differently from the He–Ne ring laser, in the Nd:YAG ring laser the simultaneous excitation of CW and CCW resonant modes is not easy to be achieved and so this kind of laser usually operates unidirectionally. This is due to the establishment of inhomogeneous saturation of the gain medium which produces a population inversion grating and so a strong coupling between the two counter-propagating modes. The introduction of an additional coupling source between CW and CCW beams counteracting the gain saturation induced coupling is necessary to realize a bi-directional Nd:YAG ring laser. In the proposed device this externally controlled coupling is due to optical loss which is different for CW and CCW signals. In particular the laser is properly designed so that the resonant mode with the highest power level suffers the largest loss.

For angular rate values larger than 100 °/s (=360,000 °/h), the fabricated devices, having a perimeter of exceeding 20 cm, is capable to sense the rotation with good linearity. In the range between 19 °/s (=68,400 °/h) and 100 °/s, a lack of linearity is observed. Finally, for $\Omega < 19$ °/s some instabilities in resonating modes intensity have been observed and no beat signal has been measured. Then a dead band extending from -19 to 19 °/s appears in the static characteristic of this gyroscope ($\Omega_L = 19$ °/s). In [46] the effect of the Nd:YAG crystal sinusoidal vibration with a frequency of 168 kHz and an amplitude of 0.47 µm on solid-state RLG performance has been investigated. This crystal vibration along the optical axis of the active cavity allows to reduce Ω_L up to 5 °/s (= 18,000 °/h).

Recently a solid-state RLG exploiting a vertical cavity surface emitting laser (VCSEL) as optical amplifier has been demonstrated [47]. The GaAs VCSEL optically pumped by a semiconductor laser diode is included in an optical cavity realized by three mirrors. Performance of this device is similar to that of the solid-state RLG based on the Nd:YAG amplifier.

References

1. Macek, W.M., Davis, D.T.M.: Rotation rate sensing with travelling-wave ring lasers. Appl. Phys. Lett. **2**, 67–68 (1963)
2. Barbour, N.: Inertial components—past, present, and future. AIAA Guidance, Navigation and Control Conference, Montreal, Canada, 6–9 August 2001
3. Schwartz, S., Feugnet, G., Pocholle, J.-P.: Diode-pumped solid-state ring laser gyroscope. Conference on Lasers and Electro-Optics/Quantum Electronics and Laser Science Conference and Photonic Applications Systems Technologies, Baltimore, USA, paper JThD47, 6–11 May 2007
4. Rosenthal, A.H.: Regenerative circulatory multiple-beam interferometry for the study of light-propagation effects. J. Opt. Soc. Am. **52**, 1143–1148 (1962)

5. Aronowitz, F.: The laser gyro. In: Ross, M. (ed.) Laser Applications. Academic Press, New York (1971)
6. Faucheux, M., Fayoux, D., Roland, J.J.: The ring laser gyro. J. Opt. **19**, 101–115 (1988)
7. Wilkinson, J.R.: Ring lasers. Progr Quantum Electron **11**, 1–103 (1987)
8. Heer, C.V.: History of the laser gyro. Proc. SPIE **487**, 2–12 (1984)
9. Killpatrick, J.: The laser gyro. IEEE. Spectr. **4**(10), 44–55 (1967)
10. Lim, W.L., Hauck, J.P., Raquet, J.W.: Pentagonal ring laser gyro design. US Patent # 4,705,398, 1987
11. Simms, G.J.: Ring laser gyroscopes. US Patent # 4,407,583, 1983.
12. Aronowitz, F.: Effects of radiation trapping on mode competition and dispersion in the ring laser. Appl. Opt. **11**, 2146–2152 (1972)
13. Aronowitz, F.: Single-isotope laser gyro. Appl. Opt. **11**, 408–412 (1972)
14. Bretenaker, F., Lépine, B., Le Calvez, A., Adam, O., Taché, J.-P., Le Floch, A.: Resonant diffraction mechanism, nonreciprocity, and lock-in in the ring-laser gyroscope. Phys. Rev. A **47**, 543–551 (1993)
15. Podgorski, T.J., Aronowitz, F.: Langmuir flow effects in the laser gyro. IEEE. J. Quantum Electron. **QE-4**, 11–18 (1968)
16. Aronowitz, F., Collins, R.J.: Mode coupling due to backscattering in a He–Ne travelling-wave ring laser. Appl. Phys. Lett. **9**, 55–58 (1966)
17. Aronowitz, F., Collins, R.J.: Lock-in and intensity-phase injection in the ring laser. J. Appl. Phys. **41**, 130–141 (1970)
18. Spreeuw, R.J.C., Neelen, R.C., van Druten, N.J., Eliel, E.R., Woerdman, J.P.: Mode coupling in a He–Ne ring laser with backscattering. Phys. Rev. A **42**, 4312–4324 (1990)
19. Kataoka, I., Kawahara, Y.: Dependence of lock-in and winking pattern on the phase-interaction of scattering waves in the ring laser. Jpn. J. Appl. Phys. **25**, 1365–1372 (1986)
20. Scully, M.O., Zubairy, M.S.: Quantum Optics. Cambridge University Press, Cambridge (1997)
21. Chow, W.W., Gea-Banacloche, J., Pedrotti, L.M., Sanders, V.E., Schleich, W., Scully, M.O.: The ring laser gyro. Rev. Mod. Phys. **57**, 61–104 (1985)
22. Thomson, A., King, P.: Ring-laser accuracy. Electron. Lett. **2**, 417 (1966)
23. Macek, W., Schneider, J., Salamon, R.: Measurement of Fresnel drag with the ring laser. J. Appl. Phys. **35**, 2556–2557 (1964)
24. Krebs, J., Maisch, W., Prinz, G., Forester, D.: Applications of magneto-optics in ring laser gyroscopes. IEEE. Trans. Magn. **16**, 1179–1184 (1980)
25. Hutchings, T., Winocur, J., Durrett, R., Jacobs, E., Zingery, W.: Amplitude and frequency characteristics of a ring laser. Phys. Rev. **152**, 467–473 (1967)
26. Andrews, D.A., King, T.A.: Sources of error and noise in a magnetic mirror gyro. IEEE J. Quantum Electron. **32**, 543–548 (1996)
27. Macek, M.: Ring laser magnetic bias mirror compensated for non-reciprocal loss. US Patent # 3,851,973, 1974
28. McClure, R.E.: Ring laser frequency biasing mechanism. US Patent # 3,927,946, 1975
29. Killpatrick, J.: Random bias for laser angular rate sensor. US Patent # 3,467,472, 1969
30. Aronowitz, F.: Fundamentals of the ring laser gyro. In: Loukianov, D., Rodloff, R., Sorg, H., Stieler, B. (eds.) Optical Gyros and their Applications. NATO Research and Technology Organization (1999)
31. Chow, W.W., Hambenne, J.B., Hutchings, T.J., Sanders, V.E., Sargent, M., Scully, M.O.: Multioscillator laser gyros. IEEE J. Quantum Electron. **QE-16**, 918–936 (1980)
32. de Lang, H.: Eigenstates of polarization in lasers. Phillips Res. Rep. **19**, 429–440 (1964)
33. Yntema, G.B., Grant, D.C., Warner, R.T.: Differential laser gyro system. US Patent # 3,862,803, 1975
34. Volk, C.H., Longstaff, I., Canfield, J.M., Gillespie, S.C.: Litton's second generation ring laser gyroscope. Proceedings of the 15th Biennial Guidance Test Symposium, Holloman Air Force Base, New Mexico, USA, pp. 493–502, 24–26 Sept 1991.

35. Sanders, V.E., Madan, S., Chow, W.W., Scully, M.O.: Beat-note sensitivity in a Zeeman laser gyro: theory and experiment. Opt. Lett. **5**, 99–101 (1980)
36. Azarova, V.V., Golyaev, Y.D., Dmitriev, V.G., Drozdov, M.S., Kazakov, A.A., Melnikov, A.V., Nazarenko, M.M., Svirin, V.N., Soloviova, T.I., Tikhmenev, N.V.: Zeeman laser gyroscopes. In: Loukianov, D., Rodloff, R., Sorg, H., Stieler, B. (eds.) Optical Gyros and their Applications. NATO Research and Technology Organization (1999)
37. Chesnoy, J.: Picosecond gyrolaser. Opt. Lett. **14**, 990–992 (1989)
38. Dennis, M.L., Diels, J.-C.M., Lai, M.: Femtosecond ring dye laser: a potential new laser gyro. Opt. Lett. **16**, 529–531 (1991)
39. Roland, J.J., Agrawal, G.P.: Optical gyroscopes. Opt. Laser Technol. **13**, 239–244 (1981)
40. Cresser, J.D., Louisell, W.H., Meystre, P., Schleich, W., Scully, M.O.: Quantum noise in ring-laser gyros. I. Theoretical formulation of the problem. Phys. Rev. A **25**, 2214–2225 (1982)
41. Schleich, W., Cha, C.-S., Cresser, J.D.: Quantum noise in a dithered-ring-laser gyroscope. Phys. Rev. A **29**, 230–238 (1984)
42. Dorschner, T.A., Haus, H.A., Holz, M., Smith, I.W., Statz, H.: Laser gyro at quantum limit. IEEE J. Quantum Electron. **QE-16**, 1376–1379 (1980)
43. Jacobs, G.B.: CO_2 laser gyro. Appl. Opt. **10**, 219–221 (1971)
44. Schwartz, S., Feugnet, G., Bouyer, P., Lariontsev, E., Aspect, A., Pocholle, J.-P.: Mode-coupling control in resonant devices: application to solid-state ring lasers. Phys. Rev. Lett. **97**, 093902 (2006)
45. Schwartz, S., Gutty, F., Pocholle, J.-P., Feugnet, G.: Solid-state laser gyro with a mechanically activated gain medium. US Patent # 0,042,225, 2008
46. Schwartz, S., Gutty, F., Feugnet, G., Loit, E., Pocholle, J.-P.: Solid-state ring laser gyro behaving like its helium-neon counterpart at low rotation rates. Opt. Lett. **34**, 3884–3886 (2009)
47. Mignot, A., Feugnet, G., Schwartz, S., Sagnes, I., Garnache, A., Fabre, C., Pocholle, J.-P.: Single-frequency external-cavity semiconductor ring-laser gyroscope. Opt. Lett. **34**, 97–99 (2009)

Chapter 4
Fiber Optic Gyroscopes

In the late 1960s, the development of the fiber optic gyroscope (FOG) was started at US Naval Research Laboratories, Washington (USA) [1]. The exploitation of optical fibers to realize an optical angular rate sensor was investigated with the hope of reducing cost, simplifying the fabrication process and increasing accuracy with respect to the He–Ne RLG. Despite a considerable research amount during the last decades, FOG has not yet superseded the He–Ne RLG because of the large existing RLG-based industrial infrastructure and the FOG higher sensitivity to external perturbations (temperature changes, vibrations and so on).

FOG performance can be very high, but also medium and low, depending on the used fiber and on the accuracy of the read-out optoelectronic system. For example, bias drift typically ranges from 10 to 0.0002 °/h. FOGs have been used in applications requiring medium and low performance such as robotics and automotive, but also in space applications, which require high or very high resolution. For example, a FOG exhibiting bias drift around 0.001 °/h has been used in the Pleiades satellites for Earth Observation [2]. Recently a FOG-based IMU has been used for inertial navigation of rovers vehicles developed by NASA for Mars exploration. Typically FOGs are not used for aircraft navigation because this application requires a high immunity to external perturbations.

High performance gyros market is actually dominated by He–Ne RLG which is widely employed for aircraft navigation but FOG has been effectively applied for inertial navigation of submarines and spacecrafts which move in a more quiet environment.

The most diffused FOGs are phase-sensitive where angular rate is estimated by measuring the rotation-induced phase shift between two beams that counter-propagate in a fiber coil. Since this phase shift is detected by an interferometric technique, these FOGs are called interferometric FOGs (IFOGs).

An optical fiber can be used to realize a passive optical ring resonator. This resonator supports a number of resonant modes. When the optical cavity is at rest, to each resonant mode corresponds a resonance frequency which is the same if the cavity is excited by an optical signal propagating either in the CW direction or in

M. N. Armenise et al., *Advances in Gyroscope Technologies*,
DOI: 10.1007/978-3-642-15494-2_4, © Springer-Verlag Berlin Heidelberg 2010

the CCW one. When the cavity rotates, to each resonant mode corresponds two resonance frequencies. One relevant to the CW propagation direction and the other to the CCW propagation direction. Difference between these two resonance frequencies is proportional to the angular rate. A FOG based on the use of a fiber ring resonator is called resonant FOG (RFOG).

An optically pumped fiber ring laser can be realized by employing either Brillouin or Raman scattering or by using an erbium-doped fiber. This fiber ring laser can be exploited as fundamental building block of an active optical gyroscope. In this device, two counter-propagating signals exhibiting a rotation-induced frequency shift are generated within the fiber ring laser. The difference between frequencies of generated signals is proportional to the angular rate.

Some reviews on FOGs are available in [3–7]. A number of papers about FOGs are collected in [8].

In this chapter operation principles of IFOG, RFOG and FOG based on a fiber ring laser are described. Then main characteristics, advantages and drawbacks of these devices are summarized.

4.1 Interferometric Fiber Optic Gyros (IFOGs)

As previously discussed, rotation induces a phase difference $\Delta\varphi$ between two counter-propagating optical signals travelling in a fiber coil, which is proportional to the angular velocity.

The basic architecture of an IFOG is shown in Fig. 4.1a [9]. A light source generates an optical signal, which is divided in two different beams by a beam splitter. The beams are coupled into the two ends of a multi-turn fiber coil by two lenses. The counter-propagating signals at the output of the two fiber ends are recombined by the beam splitter and the optical signal resulting from the interference is sent to the photodetector. The rotation induced phase difference $\Delta\varphi$ is given by:

$$\Delta\varphi = \frac{8\pi^2 R^2}{c\lambda}k\Omega = \frac{4\pi LR}{c\lambda}\Omega \tag{4.1}$$

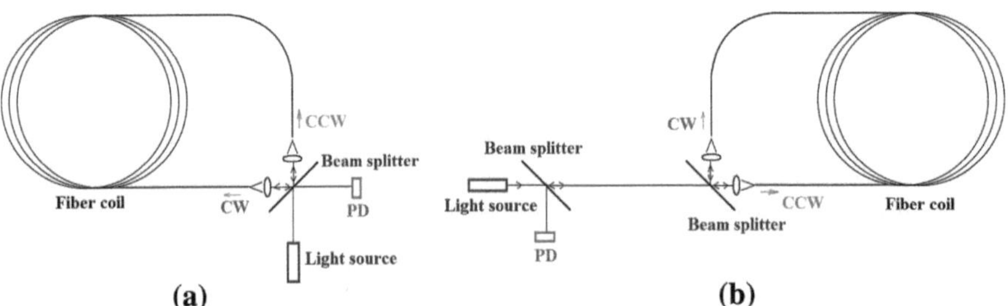

Fig. 4.1 a IFOG basic architecture. b IFOG reciprocal configuration

where λ is the counter-propagating signals wavelength (sensor operating wavelength), R is the fiber coil radius, c is the speed of light in the vacuum, k is the number of coil turns and L is the fiber total length.

Gyro scale factor depends on the optical signals wavelength, coil radius and turns number k. Then, for a given sensor size, the gyro sensitivity can be enhanced by increasing k. Unfortunately, fiber length cannot increase indefinitely because of the finite attenuation in the fiber.

Since the beam splitter induces a phase shift of $\pi/2$ between transmitted and reflected beams, the basic IFOG architecture shown in Fig. 4.1a is not reciprocal, in the sense that the two beams at the photodetector are not in phase when the gyro is at rest. The CW signal is reflected two times passing through the beam splitter whereas the CCW signals is transmitted two times through the beam splitter. The CW and CCW signals exhibit a phase shift of π at photodetector when the gyro is at rest.

The reciprocity problem can be solved employing the so-called reciprocal (or minimum) configuration shown in Fig. 4.1b [10].

In the reciprocal configuration the CW and CCW signals are both reflected and transmitted two times passing through the beam splitters so they enter the photodetector in phase when the gyro does not rotate.

As already mentioned, the optical signal at the photodetector is the result of the interference between the two counter-propagating waves. Its time-dependent power (P_{pd}) is given by the following expression:

$$P_{pd}(t) = \frac{P_{in}}{2}\{1 + \cos[\Delta\phi(t)]\} \tag{4.2}$$

where P_{in} is the optical power of the input signal generated by the light source and $\Delta\phi(t)$ is the time-dependent phase shift between the two interfering signals. If $\Delta\phi(t)$ is equal to the rotation-induced phase shift $\Delta\varphi$ (no modulation applied to CW and CCW beams), the IFOG is really low-sensitive to low angular rate values. Maximum sensitivity of the output photocurrent with respect to the angular rate is obtained when $\Delta\phi(t)$ is close to $\pm\pi/2$. By applying a phase modulation to CW and CCW signals it is possible to keep the gyro operating point where the sensitivity to angular rate is maximum. As shown in Fig. 4.2, this phase modulation can be induced by inserting an optical phase modulator after the beam splitter that produces the two counter-propagating signals [11]. This typical modulating signal is a square-wave having a period equal to $2\,\Delta\tau$ (frequency is of the order of MHz), being $\Delta\tau$ the difference between the times taken by the counter-propagating beams to propagate from the beam splitter to the phase modulator. The phase shift imposed on optical signal passing through the modulator (ϕ_m) is equal to $\pm\pi/4$ When the phase modulation is applied, $\Delta\phi(t)$ can be written as:

$$\Delta\phi(t) = \Delta\varphi + \phi_m(t) - \phi_m(t - \Delta\tau) \tag{4.3}$$

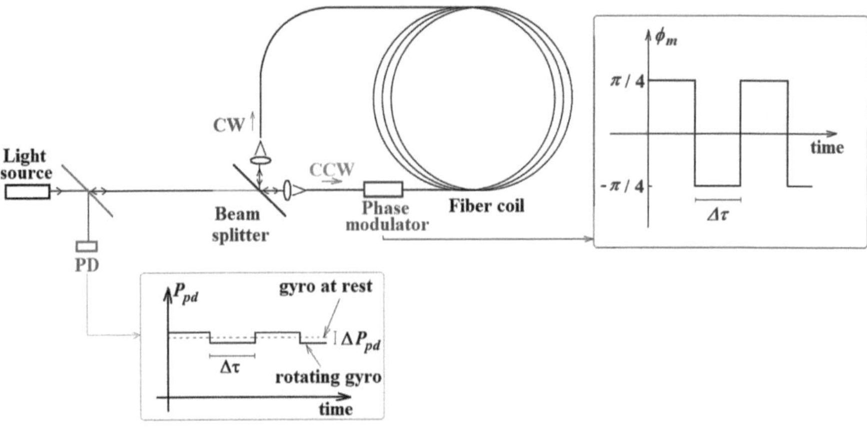

Fig. 4.2 Phase modulation of CW and CCW beams in the IFOG (open-loop configuration)

where $\phi_m(t)$ is the phase shift imposed to the CW signal and $\phi_m(t - \Delta\tau)$ is the phase shift imposed on the CCW signal. Since the period of the modulating signal is equal to $2\,\Delta\tau$, we have:

$$\Delta\phi = \Delta\varphi \pm \pi/2 \tag{4.4}$$

and so P_{pd} is equal to $P_{in}/2$ if the gyro is at rest (see Eq. 4.2). According with Eq. 4.2, when the gyro rotates, P_{pd} can be written as:

$$P_{pd}(t) = \frac{P_{in}}{2}[1 + \cos(\Delta\varphi \pm \pi/2)] = \frac{P_{in}}{2}[1 \mp \sin(\Delta\varphi)] \tag{4.5}$$

IFOG angular rate is proportional to the difference ΔP_{pd} between the two values assumed by $P_{pd}(t)$. In Fig. 4.3 the normalized ΔP_{pd} dependence on angular rate is shown for an IFOG having a fiber total length of 1 km, a fiber coil radius of 10 cm and an operating wavelength of 1.55 μm.

As we can see in Fig. 4.3, the IFOG architecture in which phase modulator is not included in a feedback loop (usually called open-loop configuration) exhibits a significant lack of linearity for quite large Ω values ($>10\ °/s = 36{,}000\ °/h$). Since a good linearity on the whole dynamic range has to be assured in high performance IFOG, open-loop configuration is typically not employed when high accuracy is required.

In closed-loop configuration shown in Fig. 4.4, the IFOG includes also a feedback loop connecting the photodetector and the phase modulator [12]. The role of this feedback loop is to generate a feedback phase shift ϕ_{fb} that is opposite to the rotation-induced phase shift $\Delta\varphi$. So, in closed-loop configuration the time-dependent phase shift between the two interfering signals is:

$$\Delta\phi(t) = \Delta\varphi + \phi_m(t) - \phi_m(t - \Delta\tau) + \phi_{fb}(t) - \phi_{fb}(t - \Delta\tau) \tag{4.6}$$

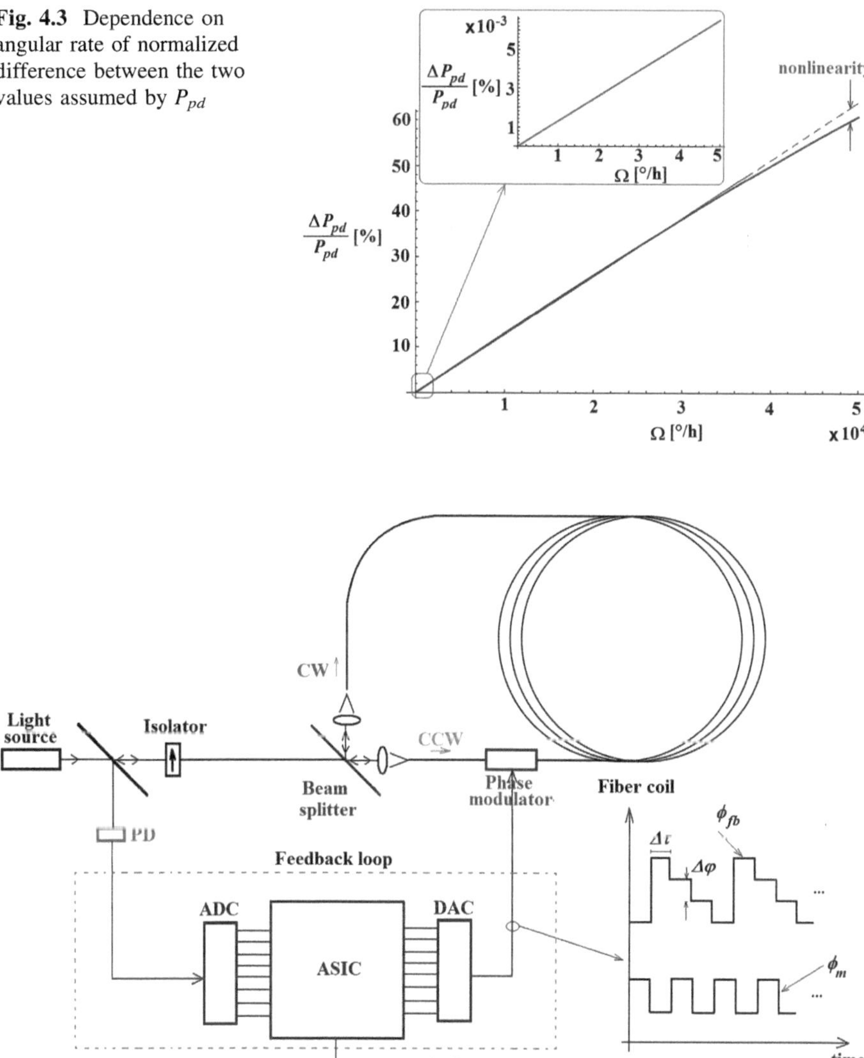

Fig. 4.3 Dependence on angular rate of normalized difference between the two values assumed by P_{pd}

Fig. 4.4 IFOG closed-loop configuration

where $\phi_{fb}(t)$ is the feedback phase shift imposed to the CW signal and $\phi_{fb}(t - \Delta\tau)$ is the feedback phase shift imposed to the CCW signal.

In recently fabricated IFOG, the two phase shifts ϕ_{fb} and ϕ_m can assume a finite set of values and the total phase shift imposed by the phase modulator is the sum $\psi_{fb} + \phi_m$. As previously pointed out, ϕ_m is a square-wave having a period equal to $2\Delta\tau$ and assuming only the values $\pm\pi/4$. The phase shift ϕ_{fb} consists of phase steps having an amplitude of $\Delta\varphi$ and a duration of $\Delta\tau$ (see the inset in Fig. 4.4).

The two phase shifts ϕ_{fb} and ϕ_m exhibit a time-dependence that fulfils the following properties:

$$\phi_{fb}(t) - \phi_{fb}(t - \Delta\tau) = -\Delta\varphi$$
$$\phi_m(t) - \phi_m(t - \Delta\tau) = \pm\pi/2$$

(4.7)

These conditions mean $\Delta\phi(t) = \pm\,\pi/2$ (see Eq. 4.6). Then the sensor operates at maximum sensitivity and the nonlinearity in the response is effectively reduced.

The feedback loop includes an analog-to-digital converter (ADC), an application-specific integrated circuit (ASIC) and a digital-to-analog converter (DAC). The ADC converts the electric signal generated by the photodetector in the digital form. The ASIC processes this electric signal generating the digital form of modulating signal and the sensor output. Finally, the DAC converts the digital form of modulating signal in the analog domain. This analog signal is sent to the phase-modulator.

IFOG performance is limited by the presence of nonreciprocal effects in the fiber coil.

Refractive index of silica depends on optical power of propagating beam. This Kerr-like nonlinearity of silica refractive index may induce a phase shift between CW and CCW beams due to the fact that the two signals may not have the same optical power [13]. This negative effect can be significantly reduced by using a broadband light source.

Silica refractive index depends also on temperature (a temperature change of 1 K induces a refractive index change of about 10^{-5}), which means that time-varying temperature gradients along the fiber coil can induce inaccuracies in the sensor response, according with Shupe effect [14]. This effect, which is enhanced by the large length of the fiber coil, can be reduced by particular winding schemes as the dipolar and quadruplar ones [15]. Vibration of the fiber coil can seriously affect the IFOG accuracy, too.

Rayleigh scattering is due to local microscopic fluctuations in silica density and may produce a power exchange between CW and CCW beams in IFOG [16]. Because of the elastic feature of Rayleigh scattering, the noise contribution due to backscattering can be effectively reduced by using a broadband light source.

Radiation-induced darkening in optical fiber consists of an increase of fiber attenuation due to the interaction between the high-energy radiation and the material forming the fiber. The amount of this effect depends on the type of the fiber, the type of radiation (gamma radiation is usually assumed to be representative), the total dose and the dose rate. In high performance IFOG the effect of radiation-induced darkening is usually compensated by controlling the power of the broadband light source which is included in a power control loop.

Polarization instabilities in CW and CCW beams degrade IFOG performance. The use of polarization maintaining fibers allows to solve this problem.

The main IFOG components are the fiber coil and the light source. The fiber coil can be realized by either single-mode telecom fiber or polarization maintaining single-mode fiber, which is commercially available but more expensive than standard telecom fiber. Polarization maintaining fiber enables to reach better accuracy and it is always used in high performance IFOGs. The employment of single-mode air-core photonic bandgap (PBG) fibers to replace conventional fibers has been recently proposed to enhance IFOG performance [17, 18]. PBG fibers, exhibiting guided modes confined in air, have been experimentally demonstrated in [19]. Because in air-core PBG fibers the most of optical power is confined in air, the Kerr and Shupe effects can be significantly reduced. This is due to the fact that refractive index of air is less sensitive to temperature and optical power of propagating beam than refractive index of silica.

The light source is broadband because the negative effect of Rayleigh backscattering and Kerr effect on IFOG accuracy can be minimized by interrogating the sensor by it. Then superluminescent diodes (emitting around 0.8 μm) or sources based on amplified spontaneous emission in erbium-doped fibers (emitting around 1.55 μm) are typically used. In high performance IFOG, broadband sources based on erbium-doped fibers are typically preferred especially because of their higher drift with respect to temperature changes.

Time-varying temperature gradients, vibrations, polarizations instabilities, Rayleigh backscattering and Kerr effect are noise sources limiting IFOG resolution. Generally the IFOG design is focused on the reduction of all noise sources below photodetector shot noise, so that IFOG can operate in the so called shot noise limited mode. In this case the minimum detectable angular rate is equal to:

$$\delta\Omega = \frac{c\lambda}{4RL}\sqrt{\frac{Bhc}{\eta\lambda P_{pd}}} \times \frac{3600 \times 180}{\pi} = \frac{c\lambda}{4RL}\Pi \times \frac{3600 \times 180}{\pi} \quad [^\circ/\text{h}] \qquad (4.8)$$

where P_{pd} is the average optical power at the input of the photodetector, η is the photodetector efficiency, B is the sensor bandwidth, h is the Planck constant and $\Pi = (B\,h\,c/\eta\,\lambda\,P_{pd})^{1/2}$ is a dimensionless parameter depending on performance of the read-out system. To enhance gyro accuracy it is necessary to increase the product $RL\sqrt{P_{pd}}$ because the minimum detectable angular rate is inversely proportional to it.

For an IFOG having a bandwidth of 20 Hz and operating at 1.55 μm, the minimum detectable angular rate dependence on the product $R \times L$ has been plotted in Fig. 4.5 for three values of P_{pd} (=0.1, 0.3, 0.5 mW). A P_{pd} increase induces a gyro sensitivity increase but a too large P_{pd} value may enhance the sensor inaccuracy due to the Kerr effect. The minimum detectable angular rate monotonically increases by increasing the $R \times L$ product. The typical value of R is around 5–10 cm whereas L, usually of the order of Km, depends on the required IFOG sensitivity. For example, if $R = 5$ cm and $P_{pd} = 0.3$ mW, the minimum detectable angular rate ranges from 0.06 to 0.015 °/h when L passes from 1 to 4 km.

In Fig. 4.6, a typical architecture of high performance IFOGs fabricated in recent years is shown. The most critical sensor components are the broadband light source, the fiber coil realized by a polarization maintaining single-mode fiber, the multifunction integrated optic chip and the feedback loop.

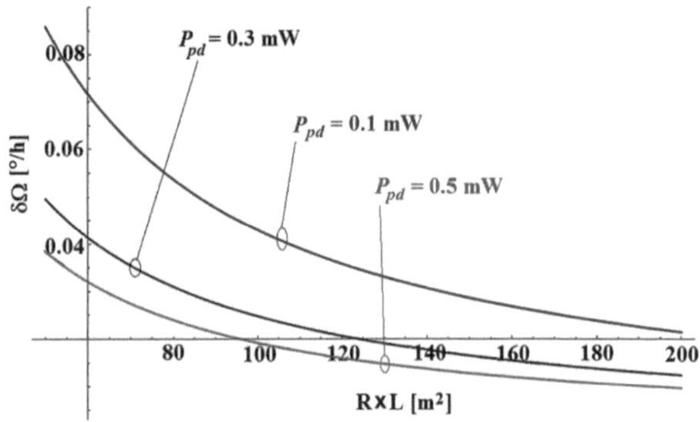

Fig. 4.5 Minimum detectable angular rate (in °/h) of IFOG as dependent on the $R \times L$ product for three values of optical power at the photodetector

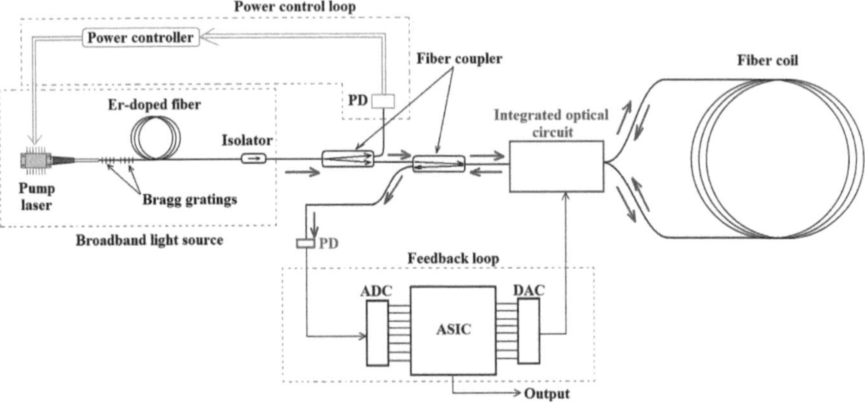

Fig. 4.6 Typical architecture of recently fabricated IFOGs [20]

The broadband light source, based on amplified spontaneous emission in an erbium-doped fiber, includes a pump laser diode operating at 980 nm, two fiber Bragg gratings reflecting the pump signal and the generated signal (centered around 1550 nm), an erbium doped fiber and an isolator. Optical power generated by the light source is adjusted by a power control loop realized by a photodetector (PD) and an electronic circuit providing the injection current to the pump laser diode.

The integrated optical chip is typically realized in LiNbO$_3$ technology and includes a polarizing waveguide, a Y-junction coupler and a phase-modulator.

An ADC, an ASIC and a DAC are used in the feedback loop generating the modulating signal.

A small fraction (1–5%) of optical signal generated by the light source is sent to the power control loop by the first fiber coupler whereas the remaining part of the optical power generated by the light source is directed toward the integrated optical circuit which splits the incoming beam in two signals. After the modulation and the propagation in the fiber coil, CW and CCW beams go back to the integrated optical circuit where they are combined. The results of the interference between CW and CCW beams is sent to the photodetector. Starting from the electrical signal generated by this photodetector, the feedback loop generates the modulating signal and the sensor output.

IFOGs performance can be very high. A bias drift less than 3×10^{-4} °/h and an angle random walk (ARW) less than 8×10^{-5} °$/\sqrt{h}$ have been achieved by a IFOG intended for space applications [20]. Unfortunately the problem of IFOG higher sensitivity with respect to He–Ne RLG to temperature changes and vibrations has not been completely solved at the moment. The dominion of He–Ne RLG over the high performance gyro market is essentially due to this IFOG drawback and to the simpler read-out electronics required by the RLG.

A very attractive research target for IFOG development may be the realization of a high performance gyro employing standard telecom fibers (not polarization maintaining). This would allow a significant IFOG cost reduction.

4.2 Resonant Fiber Optic Gyros (RFOGs)

The ring resonator is an optical component exhibiting a periodical spectral response. The period of the spectral response, called free spectral range, is the difference between two adjacent resonant frequencies. Resonance takes place when the phase-shift experienced by the optical signal propagating within the resonator is equal to an even multiple of π. When the resonator is at rest, its spectral response is independent of propagation direction (CW or CCW) of the optical signal in the cavity. When the resonator rotates, the cavity spectral responses relevant to the excitation directions (CW, CCW) suffer from a shift, which is proportional to the angular rate. While in RLG optical signals enabling the angular rate estimation are generated within the laser cavity, in the gyro based on a passive ring resonator optical signals exciting the cavity are generated outside the ring. Therefore, the RFOG is a passive optical gyro.

An optical ring resonator can be fabricated using either fiber technology or integrated optics technology. A free-space optical resonator formed by discrete optical components has been firstly proposed as sensing element of a passive gyroscope in [21]. Passive optical gyroscopes realized by integrated optics technology will be discussed in the next chapter.

The fiber ring resonator, which is the RFOG basic building block, includes a circular resonator formed by a single-mode fiber and one or two fibers to excite the resonator and to observe its spectral response. The resonator and fibers are connected by fiber couplers. The configuration including only one fiber coupler (see Fig. 4.7a) has one input port and one output port (through or reflection port). Spectral response at this port exhibits several minima corresponding to resonance frequencies. The configuration including two fiber couplers (see Fig. 4.7b) has two output ports (through or reflection port and drop or transmission port). Spectral response at drop port exhibits several maxima corresponding to resonance frequencies.

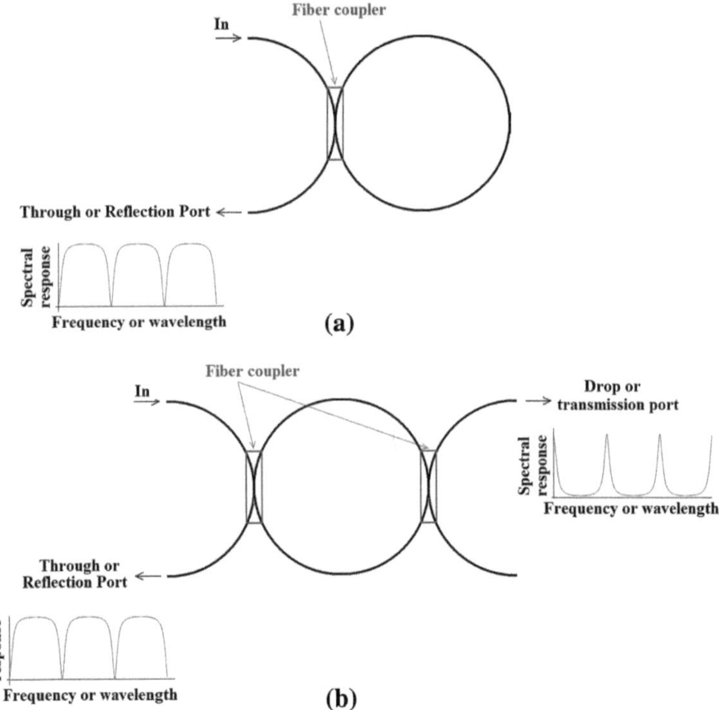

Fig. 4.7 Typical architectures of a fiber ring resonator including one (**a**) or two fiber couplers (**b**)

In a ring resonator, resonance condition is given by:

$$\beta \pi d = 2q\pi \tag{4.9}$$

where q is an integer number usually called resonance order, β is the propagation constant within the ring and d is the ring diameter. Resonance frequencies values depend on the resonance order q. When the resonator is at rest, resonance frequencies can be written as:

$$v_{q,0} = q\frac{c}{\pi d\, n_{\mathrm{eff}}} \tag{4.10}$$

where n_{eff} is the effective index of the optical mode propagating within the resonator.

As discussed in Chap. 2, rotation splits the resonance frequencies and their values are given by:

$$v_q^{CW} = q\frac{\left(c - \frac{d}{2}\Omega\right)}{\pi d\, n_{\mathrm{eff}}} \qquad v_q^{CCW} = q\frac{\left(c + \frac{d}{2}\Omega\right)}{\pi d\, n_{\mathrm{eff}}} \tag{4.11}$$

where Ω is the angular rate.

The frequency difference between two resonance frequencies having the same order (and so the same q value) and relevant to the two opposite propagation directions is equal to:

$$\Delta v = d\frac{v_{q,0}}{c}\Omega \tag{4.12}$$

From Eq. 4.11 it can be observed that the resonance frequency is independent of propagation direction of exciting optical signal if the resonator is at rest. When the resonator rotates, the resonance frequency value depends on propagation direction of the exciting beam. For a resonator formed by a multi-turn fiber coil, the difference between resonance frequency values relevant to CW and CCW excitation direction is independent of the number of coil turns and again given by the Eq. 4.12:

$$\Delta v_{\text{multi-turn}} = \Delta v = d\frac{v_{q,0}}{c}\Omega \tag{4.13}$$

Usually the number of coil turns is optimized to enhance the resonator performance, but its increase does not affect the rotation induced frequency shift between the resonance frequencies.

Of course rotation rate Ω is estimated by measuring Δv. Then a RFOG has to include a fiber resonator which is the sensing element, an optical system enabling the resonator excitation, and a read-out optoelectronic circuit allowing to estimate the frequency difference Δv.

Basically three options are possible for the operating principle of the RFOG read-out system:

- phase modulation of the optical signals exciting the fiber resonator;
- frequency modulation of the optical signals exciting the fiber resonator;
- modulation of fiber resonator length by a piezoelectric modulator.

In all these cases the use of two lock-in amplifiers (LIAs) is required. The block diagram of this electronic component is shown in Fig. 4.8. The input signal is amplified (A), pass-band filtered (PBF) and then multiplied by the reference sinusoidal signal which can be also amplified and phase shifted. Phase shifting of the reference signal can be used to cancel the phase difference between the reference signal and the input one. After the multiplication, the signal is low-pass filtered (LPF) and then amplified.

Fig. 4.8 Block diagram of the lock in amplifier

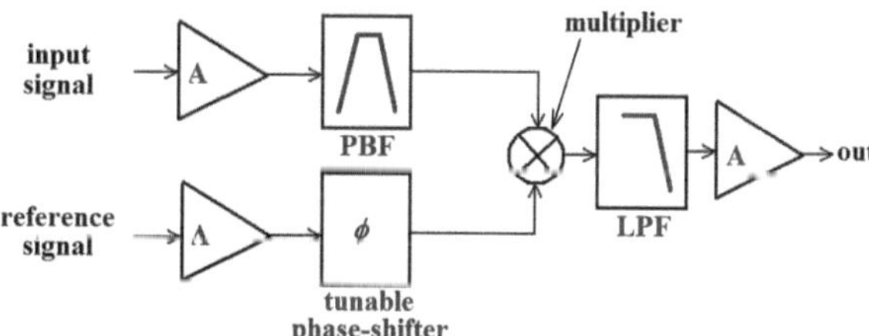

4.2.1 Phase Modulation-Based Read-Out Technique

The typical architecture for RFOG read-out based on phase modulation is shown in Fig. 4.9 [22]. This read-out technique, usually called phase modulation spectroscopy, requires the splitting of the laser beam and the phase modulation of the two resulting signals. Phase modulation is performed by two phase modulators (PM1 and PM2).

The phase modulated signals coming out from the phase modulators can be written as:

$$s_m(t) = \frac{E_0}{\sqrt{2}} e^{i[\omega_0 t + M_i \sin(\omega_m t)]} \tag{4.14}$$

where

$$M_i = \frac{\pi V}{V_\pi} \tag{4.15}$$

E_0 and ω_0 are the amplitude and the angular frequency of the laser generated signal (respectively), V and ω_m are the amplitude and the angular frequency of the sinusoidal modulating signal (respectively) and V_π is the half-wave voltage of the phase modulators PM1 and PM2. If M_i is quite near 1, the modulated signal can be expressed as:

$$s_m(t) = \frac{E_0}{\sqrt{2}} e^{i\omega_0 t} \sum_{\zeta=-2}^{2} J_\zeta(M_i) \, e^{i\zeta\omega_m t} \tag{4.16}$$

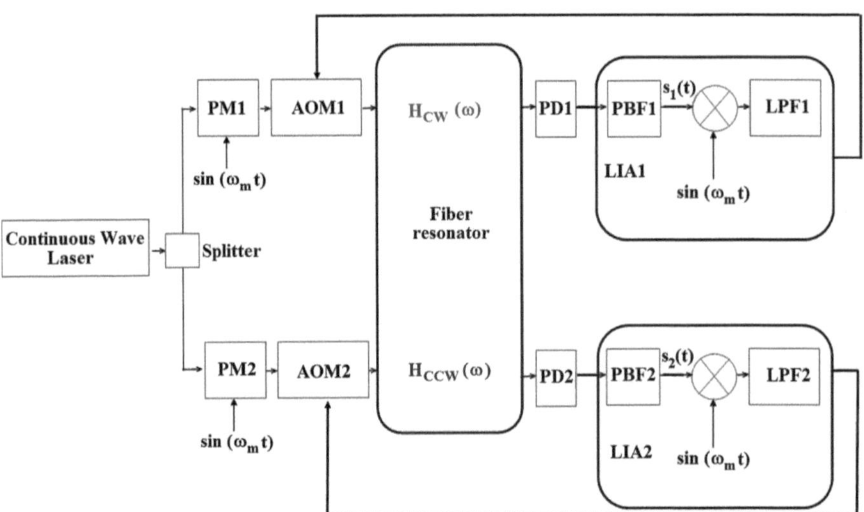

Fig. 4.9 Architecture of an RFOG read-out system using the phase modulation spectroscopy

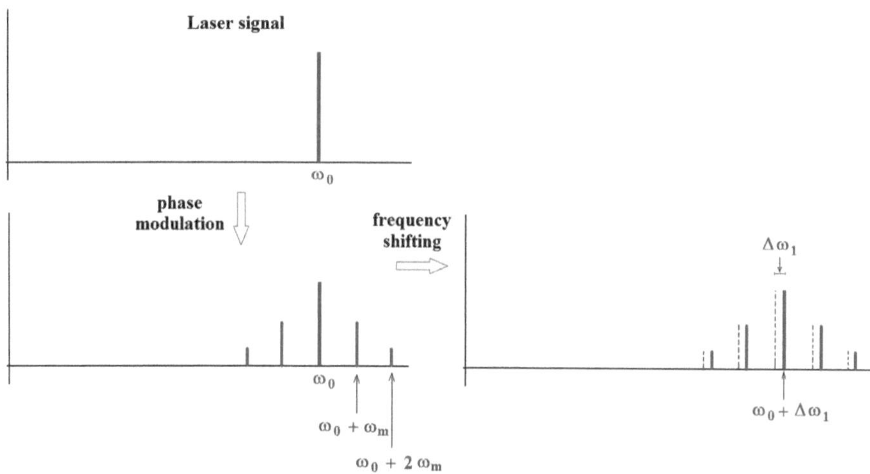

Fig. 4.10 Spectra of laser, phase modulated and frequency shifted signals

where J_ζ are Bessel functions of the first kind. According with Eq. 4.16, the phase-modulated signals have five spectral components at ω_0, $\omega_0 \pm \omega_m$ and $\omega_0 \pm 2\omega_m$.

After the phase modulation, the two signals are frequency shifted by two acousto-optic modulators (AOM1 and AOM2). The amount of the frequency shifts are equal to $\Delta\omega_1$ and $\Delta\omega_2$ for AOM1 and AOM2, respectively. Spectra of laser, phase modulated and frequency shifted signals are schematically shown in Fig. 4.10.

The signals at the output of AOM1 and AOM2 excite the fiber resonator. Spectra of these two signals are not overlapped and so power coupling between them due to backscattering is inhibited. The two signals are launched into the two ends of the fiber and coupled to the resonator. The output signals at through ports are observed (see Fig. 4.11).

Fig. 4.11 Excitation of a fiber ring resonator including only one fiber coupler by two counter-propagating signals

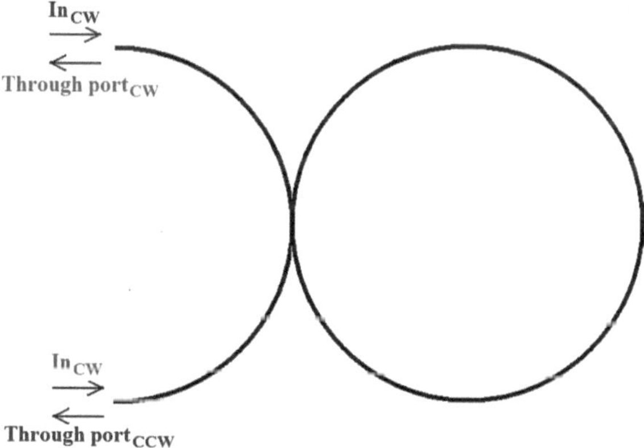

The spectral responses at the through ports are denoted $H_{CW}(\omega)$ and $H_{CCW}(\omega)$ and depend on the propagation direction (CW or CCW) of exciting signal. When the RFOG is at rest, $H_{CW}(\omega) = H_{CCW}(\omega)$.

The two signals outgoing from the through ports of the fiber resonator are sent to the two photodetectors (PD1 and PD2). The time-dependent electrical signals generated by PD1 and PD2 are:

$$s_{pd1}(t) = a_{pd1} \left| \sum_{\zeta=-2}^{2} J_\zeta(M_i)\, e^{i(\omega_0 + \Delta\omega_1 + \zeta\omega_m)t} H_{CW}(\zeta\omega_m) \right|^2 \qquad (4.17)$$

$$s_{pd1}(t) = a_{pd1} \left| \sum_{\zeta=-2}^{2} J_\zeta(M_i)\, e^{i(\omega_0 + \Delta\omega_1 + \zeta\omega_m)t} H_{CCW}(\zeta\omega_m) \right|^2 \qquad (4.18)$$

where a_{pd1} and a_{pd2} are the amplitudes of these two signals.

In the LIAs $s_{pd1}(t)$ and $s_{pd2}(t)$ are firstly pass-band filtered (the central frequency of the filter is equal to ω_m) obtaining two sinusoids whose frequency is ω_m. Since every sinusoid can be expressed as the sum of a sine function (in-phase component) and a cosine function (quadrature component), signals at the output of the pass-band filters included in the two LIAs can be written as:

$$
\begin{aligned}
s_1(t) &= A_1 \cos(\omega_m t) + B_1 \sin(\omega_m t) \\
s_2(t) &= A_2 \cos(\omega_m t) + B_2 \sin(\omega_m t)
\end{aligned}
\qquad (4.19)
$$

As proved in [23], the amplitudes of quadrature components of these pass-band signals (B_1 and B_2) are proportional to the differences:

$$
\begin{aligned}
\frac{\omega_0 + \Delta\omega_1}{2\pi} - v_q^{CW} \\
\frac{\omega_0 + \Delta\omega_2}{2\pi} - v_q^{CCW}
\end{aligned}
\qquad (4.20)
$$

where v_q^{CW} and v_q^{CCW} are the two resonance frequencies of the fiber resonator relevant to the CW and the CCW propagation direction, respectively. In Fig. 4.12 the typical B_1 dependence on the difference $(\omega_0 + \Delta\omega_1)/2\pi - v_q^{CW}$ is shown. If this difference is in the range between ± 10 kHz, B_1 is directly proportional to $(\omega_0 + \Delta\omega_1)/2\pi - v_q^{CW}$. The dependence of B_2 on $(\omega_0 + \Delta\omega_2)/2\pi - v_q^{CCW}$ is similar to that of B_1 on $(\omega_0 + \Delta\omega_1)/2\pi - v_q^{CW}$.

The two LIAs allow the extraction of the amplitude of the quadrature components B_1 and B_2. In fact in the LIAs, $s_1(t)$ and $s_2(t)$ are multiplied by the modulating sinusoidal signal having the frequency equal to ω_m and then low-pass filtered. DC signals at the output of the LIAs can be used as error signals in two feedback loops, which provide the two electrical signals driving AOM1 and AOM2. In this manner feedback loops lock $(\omega_0 + \Delta\omega_1)/2\pi$ to v_{CW} and $(\omega_0 + \Delta\omega_2)/2\pi$ to v_{CCW}. When this locking is achieved, $\Delta\omega_1 - \Delta\omega_2$ is proportional to the angular rate.

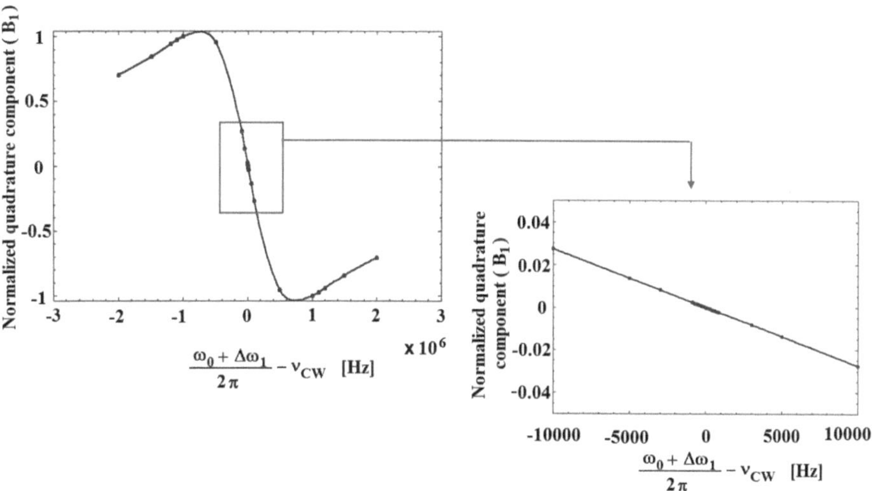

Fig. 4.12 Normalized quadrature component (B_I) versus $[(\omega_0 + \Delta\omega_1)/2\pi - v_q^{CW}]$

The read-out architecture based on phase modulation spectroscopy can also be modified with respect to that shown in Fig. 4.9. In particular the frequency shifting can be obtained by tuning the laser. In this case AOM1 and AOM2 are not necessary [24]. Moreover the modulating electric signal driving PM1 and PM2 may have either a digital serrodyne or a triangle waveform [25, 26].

4.2.2 Frequency Modulation-Based Read-Out Technique

An alternative to the phase modulation of optical signals exciting the fiber resonator is the frequency modulation of the laser beam. In this case, the read-out technique is called frequency modulation spectroscopy. The typical architecture for RFOG read-out based on frequency modulation spectroscopy is shown in Fig. 4.13 [27].

Assuming for the modulating signal a cosinusoidal time-dependence, the expression of the laser signal is:

$$s_m(t) = \frac{E_0}{\sqrt{2}} e^{i\left[\omega_0 t + M_f \int_{-\infty}^{t} \cos(\omega_m t)\right]} \tag{4.21}$$

where E_0 and ω_0 are the amplitude and the central frequency of the laser beam, respectively, ω_m is the angular frequency of the modulating signal, and M_f is the ratio between frequency deviation imposed by the frequency modulation and $\omega_m/2\pi$.

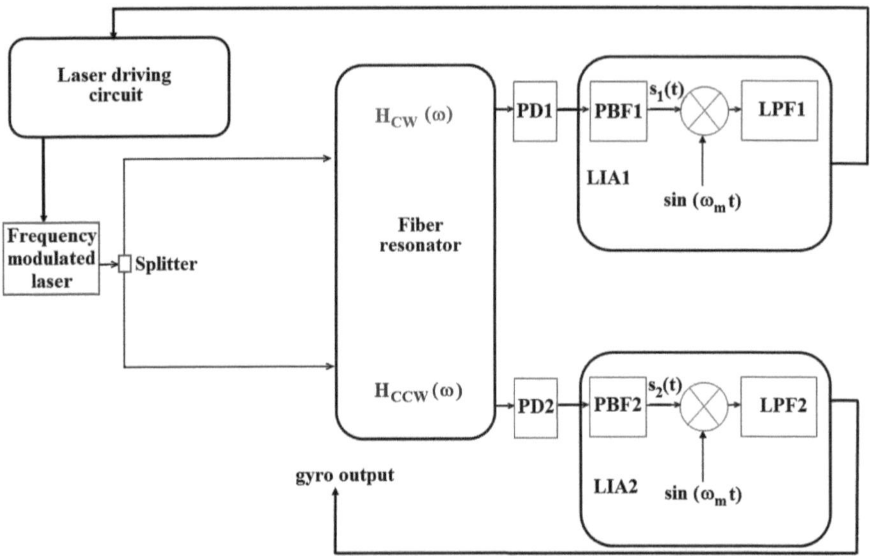

Fig. 4.13 Architecture for RFOG read-out by frequency modulation spectroscopy

Since it results:

$$\frac{E_0}{\sqrt{2}}e^{i\left[\omega_0 t + M_f \int_{-\infty}^{t} \cos(\omega_m t)\right]} = \frac{E_0}{\sqrt{2}}e^{i\left[\omega_0 t + M_f \sin(\omega_m t)\right]} \tag{4.22}$$

the mathematical description of this read-out technique is the same of the technique based on the phase modulation spectroscopy. The DC signal outgoing from LIA1 is proportional to $(\omega_0/2\pi - v_q^{CW})$ and it is used as error signal in the feedback loop, which provides the electric signal driving the laser. This feedback loop tends to lock $\omega_0/2\pi$ to v_q^{CW}. When this locking is achieved, the LIA2 output is proportional to the difference between v_q^{CW} and v_q^{CCW}. This difference is proportional to the angular rate.

The use of this read-out technique requires a very sophisticated laser source providing a stable frequency modulated optical beam with an appropriate frequency resolution.

4.2.3 Resonator Length Modulation-Based Read-Out Technique

Sinusoidal modulation of the fiber resonator length enables the sinusoidal modulation of resonance frequencies relevant to the two propagation directions [28]. In this case, the two resonance frequencies exhibit a time-dependence described by following relations:

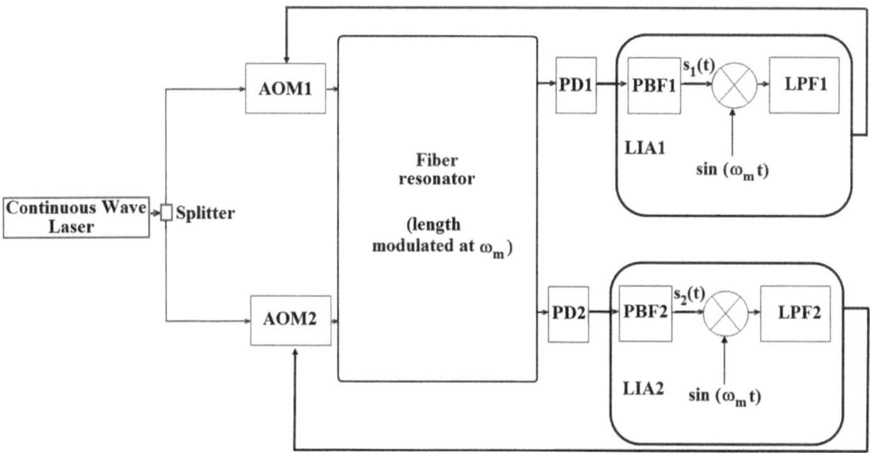

Fig. 4.14   Architecture of the RFOG read-out by resonator length modulation

$$v_{q,m}^{CW}(t) = v_q^{CW} + a_m \sin(\omega_m t)$$
$$v_{q,m}^{CCW}(t) = v_q^{CW} + a_m \sin(\omega_m t)$$

(4.23)

where a_m is the amplitude of resonance frequencies shift and ω_m is the angular frequency of modulating signal.

The architecture of this technique is shown in Fig. 4.14. The modulating signal, having a frequency of a few kHz, can be applied to a piezoelectric transducer having a cylindric shape. A portion of the fiber forming the resonator is wrapped around the piezoelectric cylinder and so that resonator length can be modulated. This induces the modulation of the two resonant frequencies.

The signals exciting the resonator are frequency shifted by two acousto-optic modulators (AOM1 and AOM2). Frequencies of exciting signals are equal to $(\omega_0 + \Delta\omega_1)$ and $(\omega_0 + \Delta\omega_2)$. The DC signal coming out from LIA1 is proportional to the difference $(\omega_0 + \Delta\omega_1)/2\pi - v_q^{CW}$. Similarly the DC signal coming out from LIA2 is proportional to the difference $(\omega_0 + \Delta\omega_2)/2\pi - v_q^{CCW}$. The two feedback loops including LIA1 and LIA2 lock $(\omega_0 + \Delta\omega_1)/2\pi$ to v_q^{CW} and $(\omega_0 + \Delta\omega_2)/2\pi$ to v_q^{CCW}. Then, after locking achievement, the angular rate is proportional to $\Delta\omega_1 - \Delta\omega_2$.

The main drawback of this read-out technique is due to the presence of the piezoelectric transducer within the resonator. This generates a significant loss contribution within the fiber resonator. This loss contribution limits the resonator performance and the RFOG overall accuracy.

4.2.4 Critical Aspects of RFOGs

Research interest for RFOGs is eminently due to the fact that theoretically a RFOG has the same performance of an IFOG but it requires a significantly shorter fiber length. For this reason RFOG is expected to be less sensitive than IFOG to vibrations and temperature gradients. On the other side, in RFOG it is more difficult to limit negative effect on accuracy due to Kerr-like nonlinearity of silica refractive index because in RFOG is not possible to use a broadband light source as in IFOG. The only way to limit the inaccuracy due to the Kerr effect is to reduce optical power of signals propagating within the fiber resonator.

Moreover the read-out system of a high performance RFOG (resolution less then 0.1 °/h) has to include an expensive laser having a linewidth less than 10 kHz. This is due to the fact that the optical signal utilized to measure the angular rate has to exhibit a full-wave half-maximum (FWHM) which must be significantly less than the cavity FWHM.

The shot noise-limited minimum detectable angular rate is given by:

$$\delta\Omega = \delta v \frac{\lambda}{d} \sqrt{2}\, \Pi \, \times \, \frac{3600 \times 180}{\pi} \quad [°/\text{h}] \tag{4.24}$$

where δv is the cavity resonance FWHM. In a resonator realized by a multi-turn fiber coil, resolution dependence on the number of coil turns is due to the fact that a turn number optimization can produce a δv reduction.

Assuming a fiber ring resonator having a diameter of 10 cm, the $\delta\Omega$ dependence on δv has been plotted in Fig. 4.15. In this plot, the average power at the input of the photodetector has been assumed equal to 0.1 mW. To realize a RFOG having high performance ($\delta\Omega < 0.1$ °/h, bias drift < 0.01 °/h) it is required $\delta v < 0.1$ MHz, which can be achieved by:

- the use of low-loss optical fibers because δv strongly depends on propagation loss within the cavity;
- the use of a multi-turn resonator having an optimized number of turn so that δv is minimized;
- the use of a laser source having a very narrow linewidth for the resonator excitation because the laser linewidth has to be significantly less than δv (laser linewidth has to be in the range 5–10 kHz);

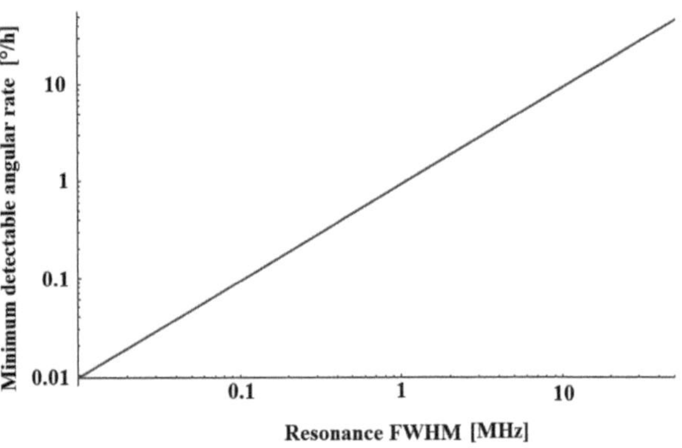

Fig. 4.15 RFOG resolution as a function of the resonator FWHM (log–log scale)

- an accurate control of power difference between CW and CCW beams to assure that bias drift due to Kerr effect does not degrade the RFOG performance.

Although RFOG with a short length resonator can theoretically achieve performance similar to that one of IFOG (typical fiber total length is around 5–10 m in RFOG and 1–5 km in IFOG), a high performance RFOG is surely more expensive than a high performance IFOG especially because of the very narrow laser linewidth required. Moreover bias drift experimentally achieved by RFOG is around 0.5 °/h, which is much higher than the value obtainable by IFOG (around 0.001 °/h).

The use of air-core photonic bandgap fibers for the RFOG realization is currently considered as an attractive research target because this kind of fibers are less sensitive to the Kerr effect than standard fibers [29].

4.3 Optical Gyros Based on a Fiber Ring Laser

An optically pumped fiber ring laser, in which two counter-propagating laser beams are generated, can be used to sense rotation. This is due to the fact that, when the ring rotates, the two counter-propagating signals generated within the laser exhibit a frequency shift, which is proportional to the angular rate. If these two signals interfere, they produce a beat signal whose amplitude oscillates with a frequency equal to the rotation-induced frequency shift. The beat signal is sent to a photodetector, which generates a photocurrent having a sinusoidal time-dependence. The frequency of this electrical signal is proportional to the angular rate. The operating principle of this kind of gyroscope is similar to that one of He–Ne RLG.

The main problem in the design of fiber ring lasers for gyro applications is the mode competition that can prevent the laser bidirectionality. Fiber ring lasers for rotation sensing are typically based on stimulated Brillouin scattering (SBS) [30, 31]. On the basis of this physical phenomenon an optical signal (pump signal) having a power larger than a threshold value (around 5–10 mW) and propagating within an optical fiber generates a signal (Stokes signal) propagating in opposite direction with respect to the propagation direction of the pump signal. For a pump wavelength of 1.55 μm, the frequency of the Stokes signal is around 11 GHz less than the frequency of the pump signal [32].

If two pump signals having a sufficiently large power are coupled within a fiber ring, two Stokes signals are generated within the ring laser due to SBS. The two Stokes signals experience a rotation-induced frequency shift, which is proportional to the angular rate. If Stokes signals interfere, a beat signal can be observed. The amplitude of this beat signal oscillates with a frequency equal to the rotation-induced frequency shift.

The basic architecture of a FOG based on SBS (called Brillouin FOG, BFOG) is shown in Fig. 4.16. The laser beam generated by the pump laser is split in two beams (P1 and P2) having the same frequency. As the pump signals propagate in

Fig. 4.16 BFOG basic
architecture

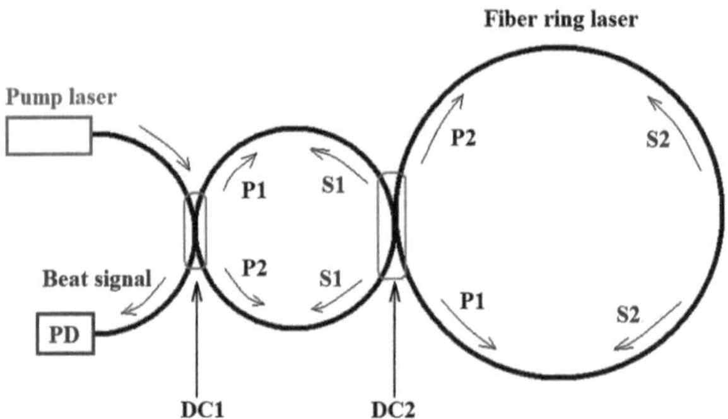

the fiber ring, the two Stokes signals (S1 and S2) are excited. A portion of these two signals is extracted from the ring through DC2 and the resulting signals interfere in the directional coupler DC1. The beat signal is sent to a photodetector (PD) and electric signal coming out from it has a frequency which is proportional to the angular rate.

The main advantage of BFOG with respect to other FOGs is the simpler readout system. On the other side, the BFOG suffers from the problem of lock-in caused by backscattering of light inside the resonator. Some optical dithering techniques to reduce the lock-in effect in the BFOG have been proposed in [33, 34].

A fiber ring laser can be realized also by exploiting the stimulated Raman scattering (SRS) [32]. In this case the pumping signal is required to have a very high power to induce the generation of Stokes signals (threshold power is around 500–600 mW). For SRS the frequency difference between the pump and the Stokes signal is around 13 THz [32]. SRS can induce Stokes signals either in the same direction of pump signal or in the opposite direction. So it is quite complex to realize a fiber ring laser suitable for rotation sensing by exploiting SRS. In [35], an active FOG based on SRS has been reported. In this gyro the fiber ring laser is optically pumped by high power pulses generated by a mode-locked laser.

In principle, also erbium-doped fiber ring lasers can be used for rotation sensing [36, 37]. Gain competition between the two counter-propagating signals can again prevent the laser bidirectionality. Moreover it is quite difficult to obtain stable operation of the laser in single longitudinal mode for both CW and CCW directions. Recently, a fiber ring laser having a quite complex architecture including circulators, fiber Bragg gratings and erbium doped fibers has been proposed to sense rotation [38]. The gyro performance of this system has not yet been evaluated.

A mode-locked fiber laser has been also proposed for rotation sensing [39]. The laser cavity is formed by a planar mirror at one end and a Sagnac interferometer (acting as a loop reflector) at the other end, with an erbium-doped fiber amplifier in between. A phase modulator is inserted within the Sagnac loop to avoid the lock-in. The output of the proposed gyroscope is a train of optical pulses. Time interval between two consecutive pulses linearly depends on the rotation rate and can be measured by a conventional electronic counter or a phase sensitive detection technique employing a lock-in amplifier. The experimentally measured bias drift of this fiber gyroscope is around 20 °/h [40]. This value of bias drift is several orders of magnitude larger than that exhibited by IFOGs and it is too large for some applications.

We believe that, despite their attractiveness due to the quite simple read-out, BFOGs and in general all FOGs based on an optically pumped fiber ring laser are not a near-term alternative to the IFOG.

References

1. Brown, R.B.: NRL Memorandum Report N1871. Naval Research Lab., Washington (1968)
2. Buret, T., Ramecourt, D., Honthaas, J., Paturel, Y., Willemenot, E., Gaiffe, T.: Fibre optic gyroscopes for space application. In: Optical Fiber Sensors, Cancún, Mexico, paper MC4, 23–27 October 2006
3. Culshaw, B., Giles, I.P.: Fiber optic gyroscopes. J. Phys. E Sci. Instrum. **16**, 5–15 (1983)
4. Bergh, R.A., Lefèvre, H.C., Shaw, H.J.: An overview of fiber-optic gyroscopes. J. Lightwave Technol. **LT-2**, 91–107 (1984)
5. Lefèvre, H.: The Fiber-Optic Gyroscope. Artech House, Norwood (1993)
6. Lefèvre, H.: Application of the Sagnac effect in the interferometric fiber-optic gyroscope. In: Loukianov, D., Rodloff, R., Sorg, H., Stieler, B. (eds.) Optical Gyros and their Applications. NATO Research and Technology Organization, France (1999)
7. Culshaw, B.: The optical fiber Sagnac interferometer: an overview of its principles and applications. Meas. Sci. Technol. **17**, R1–R16 (2006)
8. Smith, R.B. (ed.): Selected Papers on Fiber Optic Gyroscopes. SPIE Milestone Series (MS 8). Bellingham, SPIE Optical Engineering Press, Washington (1989)
9. Vali, V., Shorthill, R.W.: Fiber ring interferometer. Appl. Opt. **15**, 1099–1100 (1976)
10. Ulrich, R.: Fiber-optic rotation sensing with low drift. Optics Lett. **5**, 173 175 (1980)
11. Martin, J.M., Winkler, J.T.: Fiber-optic laser gyro signal detection and processing technique. Proc. SPIE **139**, 98–102 (1978)
12. Davis, J.L., Ezekiel, S.: Closed-loop, low-noise fiber-optic rotation sensor. Optics Lett. **6**, 505–507 (1982)
13. Ezekiel, S., Davis, J.L., Hellwarth, R.W.: Intensity dependent nonreciprocal phase shift in a fiberoptic gyroscope. In: Springer Series in Optical Sciences, Vol. 32, pp. 332–336. (1982)
14. Shupe, D.M.: Fiber resonator gyroscope: sensitivity and thermal nonreciprocity. Appl. Opt. **20**, 286–289 (1981)
15. Frigo, N.J.: Compensation of linear sources of non-reciprocity in Sagnac interferometers. Proc. SPIE **412**, 268–271 (1993)
16. Cutler, C.C., Newton, S.A., Shaw, H.J.: Limitation of rotating sensing by scattering. Optics Lett. **5**, 488–490 (1980)
17. Kim, H.K., Digonnet, M.J.F., Kino, G.S.: Air-core photonic-bandgap fiber-optic gyroscope. J. Lightwave Technol. **24**, 3169–3174 (2006)
18. Blin, S., Kim, H.K., Digonnet, M.J.F., Kino, G.S.: Reduced thermal sensitivity of a fiber-optic gyroscope using an air-core photonic-bandgap fiber. J. Lightwave Technol. **25**, 861–865 (2007)
19. Cregan, R.F., Mangan, B.J., Knight, J.C., Birks, T.A., Russell, P.S.J., Roberts, P.J., Allan, D.C.: Single-mode photonic band gap guidance of light in air. Science **285**, 1537–1539 (1999)
20. Divakaruni, S., Sanders, S.: Fiber optic gyros: a compelling choice for high precision applications. In: Optical Fiber Sensors, Cancún, Mexico, paper MC2, 23–27 October 2006

21. Ezekiel, S., Balsamo, S.R.: Passive ring resonator laser gyroscope. Appl. Phys. Lett. **30**, 478–480 (1977)
22. Zhang, X., Ma, H., Zhou, K., Jin, Z.: Experiments by PM spectroscopy in resonator fiber optic gyro. Opt. Fiber Technol. **13**, 135–138 (2007)
23. Carroll, R., Coccoli, C.D., Cardarelli, D., Coate, G.T.: The passive resonator fiber optic gyro and comparison to the interferometer fiber gyro. Proc. SPIE **719**, 169–177 (1986)
24. Zhang, X., Ma, H., Jin, Z., Ding, C.: Open-loop operation experiments in a resonator fiber-optic gyro using the phase modulation spectroscopy technique. Appl. Opt. **45**, 7961–7965 (2006)
25. Hotate, K., Harumoto, M.: Resonator fiber optic gyro using digital serrodyne modulation. J. Lightwave Technol. **15**, 466–473 (1997)
26. Jin, Z., Yang, Z., Ma, H., Ying, D.: Open-loop experiments in a resonator fiber-optic gyro using digital triangle wave phase modulation. IEEE Photonic Technol. Lett. **19**, 1685–1687 (2007)
27. Imai, T., Nishide, K.-I., Ochi, H., Ohtsu, M.: The passive ring resonator fiber optic gyro using modulatable highly coherent laser diode module. Proc. SPIE **1585**, 153–162 (1992)
28. Meyer, R.E., Ezekiel, S., Stowe, D.W., Tekippe, V.J.: Passive fiber-optic ring resonator for rotation sensing. Optics Lett. **8**, 644–646 (1983)
29. Sanders, G.A., Strandjord, L.K., Qiu, T.: Hollow core fiber optic ring resonator for rotation sensing. Optical Fiber Sensors, Cancún, Mexico, paper ME6, 23–27 October 2006
30. Thomas, P.J., van Driel, H.M., Stegeman, G.I.A.: Possibility of using an optical fiber Brillouin ring laser for inertial sensing. Appl. Opt. **19**, 1906–1908 (1980)
31. Stokes, L.F., Chodorow, M., Shaw, H.J.: All-fiber stimulated Brillouin ring laser with submilliwatt pump threshold. Optics Lett. **7**, 509–511 (1982)
32. Argawal, G.P.: Fiber-Optic Communication Systems. Wiley-Interscience, New York (2002)
33. Huang, S., Toyama, K., Kim, B.Y., Shaw, H.J.: Lock-in reduction technique for fiber-optic ring laser gyros. Optics Lett. **18**, 555–557 (1993)
34. Zarinetchi, F., Smith, S.P., Ezekiel, S.: Stimulated Brillouin fiber-optic laser gyroscope. Optics Lett. **16**, 229–231 (1991)
35. Nakazawa, M.: Synchronously pumped fiber Raman gyroscope. Optics Lett. **10**, 193–195 (1985)
36. Kim, S.K., Kim, H.K., Kim, B.Y.: Er^{3+}-doped fiber ring laser for gyroscope applications. Optics Lett. **19**, 1810–1812 (1994)
37. Kiyan, R., Kim, S.K., Kim, B.Y.: Bidirectional single-mode Er-doped fiber-ring laser. IEEE Photonics Technol. Lett. **8**, 1624–1626 (1996)
38. Lu, J., Chen, S., Bai, Y.: Experimental study on a novel structure of fiber ring laser gyroscope. Proc. SPIE **5634**, 338–342 (2005)
39. Jeon, M.Y., Jeong, H.J., Kim, B.Y.: Mode-locked fiber laser gyroscope. Optics Lett. **18**, 320–322 (1993)
40. Hong, J.B., Yeo, Y.B., Lee, B.W., Kim, B.Y.: Phase sensitive detection for mode-locked fiber laser gyroscope. IEEE Photonics Technol. Lett. **11**, 1030–1032 (1999)

Chapter 5
Integrated Optical Gyroscopes

Integrated optics has led to the development of miniaturized optical devices that have very complex functionality on a single chip. A number of integrated optical devices, such as lasers, amplifiers, multiplexers/demultiplexers, filters, modulators and switches have been fabricated in a variety of substrate materials, including crystals, glass, polymers, and semiconductors. In the last few years a considerably research effort has been spent to design and fabricate complex photonic integrated circuits (PICs) in which a number of optical components has been integrated on a single chip. For example, an InP-based PIC including more than 50 optical components has been demonstrated in [1].

Fabrication of optical gyros by integrated optics technology is a very attractive research target because it enables weight and dimension reduction, cost lowering, decrease of power consumption, better control of thermal effects and reliability increase, together with potential of full integration of the optical gyroscope system.

In active optical gyros, two resonant modes are excited within a ring laser and they experience a rotation-induced frequency shift that can be measured by an interferometric technique.

Passive optical angular rate sensors can be phase sensitive or frequency sensitive. In frequency sensitive gyros two resonance frequencies of an optical cavity relevant to clockwise and counter-clockwise propagation directions are measured. In phase sensitive gyros the rotation-induced phase shift between two beams counter-propagating in a ring interferometer is measured. Both active and passive integrated optical gyros have been proposed and fabricated. Most of passive integrated optical gyros are frequency sensitive, but slow-light phase sensitive integrated optical gyros have been recently proposed.

In this chapter the state-of-the-art of active and passive integrated optical gyros is reviewed. A wide spectrum of materials used for the rotation sensors is considered, including indium phosphide, gallium arsenide, silicon, silicon oxide, glass and lithium niobate. Achievements and future research objectives related to the fabrication of PICs for angular rate sensing are discussed.

M. N. Armenise et al., *Advances in Gyroscope Technologies*,
DOI: 10.1007/978-3-642-15494-2_5, © Springer-Verlag Berlin Heidelberg 2010

5.1 Active Integrated Optical Gyros

The sensing element of active integrated optical gyros is an integrated ring laser in which two counter-propagating beams are generated. The read-out optoelectronic circuit, preferably integrated on the same substrate on which the ring laser is fabricated, measures the rotation-induced splitting in frequency of these resonant modes.

Minimum detectable angular rate of active integrated optical gyros is limited by quantum noise and it is given by:

$$\delta\Omega = \frac{\delta v}{S}\Pi \times \frac{180 \times 3600}{\pi} \, (°/h) \tag{5.1}$$

where S is the gyro scale factor (see Eq. 3.3), δv is the linewidth of lasing emission, and Π is the parameter taking into account readout system performance (see Eq. 4.8).

The angle random walk induced by quantum noise is given by:

$$W_{ARW} = \frac{\delta\Omega}{60\sqrt{B}} = \frac{\delta v}{S\sqrt{B}}\Pi \times \frac{180 \times 60}{\pi} \left(°/\sqrt{h}\right) \tag{5.2}$$

where $\delta\Omega$ is expressed in °/h and B is the sensor bandwidth.

Scale factor maximization and laser linewidth minimization are required for improving the resolution and enhancing the ARW. An integrated ring laser to be used in high performance optical gyroscope has to exhibit a diameter exceeding a few millimetres and a linewidth less than 1 MHz.

5.1.1 Integrated Ring Lasers

The integrated ring laser is the basic building block in all active integrated optical gyros and its performance heavily affects the minimum detectable angular rate of the rotation sensor.

In the last decades integrated ring lasers have been widely investigated because they can be employed not only for rotation sensing but also in optical telecommunications and all-optical signal processing.

Electrically pumped integrated ring lasers can be realized only by III–V semiconductors technology whereas for optically pumped integrated ring lasers fabrication some different materials have been used such as indium phosphide, silicon, polymers and lithium niobate.

5.1.1.1 Electrically Pumped Integrated Ring Lasers

Performance of electrically pumped semiconductor ring lasers (SRLs), demonstrated in 1980 for the first time [2], has been significantly enhanced during the last decades. Several resonant cavity shapes have been investigated, e.g. triangular, squared, circular and racetrack (formed by two parallel straight waveguides connected by two semicircular guiding structures).

In Fig. 5.1 the first SRL proposed in literature is sketched, where the active region is a double AlGaAs/GaAs heterostructure [2, 3]. The disk-shaped (radius around several tens of microns) optical resonator is out-coupled by a Y-junction. Main drawbacks of this ring laser are its multimodal behaviour and the too high threshold current (around 200 mA).

Continuous wave and single longitudinal mode operation have been firstly reported in [4, 5], respectively. Also for these lasers a circular-shaped optical cavity and a Y-junction output coupler have been exploited. They exhibit a relatively low threshold current (less than 100 mA) with respect to the above mentioned devices.

Although the large number of SRLs that have been designed and experimented in the last 25 years, only a few devices have a cavity total length exceeding a few millimeters and are suitable for fabricating active integrated optical gyros.

Disk, triangular and squared SRLs have usually a total length which does not exceed some hundreds of microns because they have been designed for applications related to all-optical signal processing and switching, which do not require a laser cavity length in the range of millimetres [6–8].

Recently a micro-disk laser in III–V semiconductors technology has been realized on a Silicon-on-Insulator (SOI) substrate [9]. Although laser dimensions do not enable it to be used in for gyro applications, the heterogeneous integration by molecular bonding of a SRL in InGaAsP/InP and other electrical and optical components realized on a SOI substrate may be interesting for the gyro complete integration on a single chip.

First large dimensions (total cavity length around 10 mm) SRL fabricated in InGaAsP/InP has been reported in [10]. This device employs a ring resonator in which two 0.5 mm long sections are active and the remaining part of the cavity is

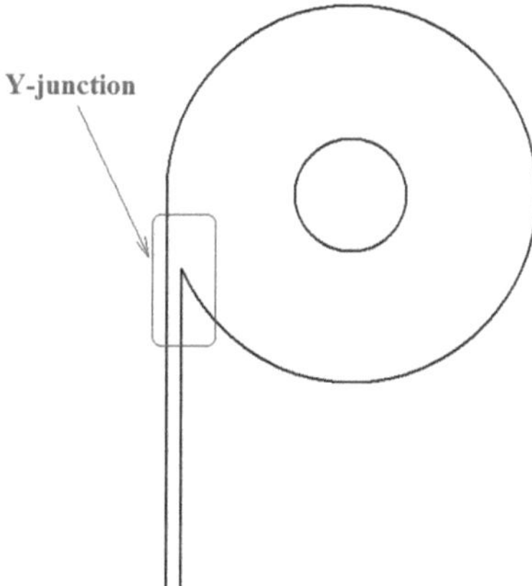

Fig. 5.1 Circular SRL exploiting a Y-junction as output coupler [3]

passive. The laser emits at 1540 nm and the linewidth of generated signal is equal to 900 kHz.

A better performing large radius SRL in AlGaAs/GaAs has been reported in [11]. This device, having a radius of 1 mm, includes a ring cavity weakly coupled (coupling efficiency in the range 1–5%) to a straight output waveguide (see Fig. 5.2a). To reduce back-reflections, the output waveguide has a 5° tilting angle with respect to cleaved facets of the substrate. Two separate metal contacts have been deposited on the straight waveguide ends acting as photodetectors (they are reversed biased at 1.5 V). Guiding structure (Fig. 5.2b) is a shallow-etched (etch depth ~ 1 μm) ridge waveguide fabricated by e-beam lithography and reactive ion etching. A double quantum well (DQW) structure grown by metal-organic chemical vapour deposition (MOCVD) has been used as active region. Laser threshold current is heavily depending on the waveguide etch depth and minimum threshold current value has been achieved for an etch depth equal to 950 nm. For an etch depth of 1050 nm, a threshold current around 270 mA has been measured.

Operating regime and linewidth of this SRL have been investigated in [11–13]. The laser exhibits three operating regimes, bidirectional continuous-wave (*bi-cw*), bidirectional with alternate oscillations (*bi-ao*) and unidirectional (*uni*) regime. When injection current is just above threshold current, two counter-propagating modes operating in continuous wave regime are excited. With the injection current in the range from 360 to 480 mA, two counter-propagating resonant modes modulated by sinusoidal oscillations have been observed. Finally, when injection current is larger than a specific value, 480 mA, the laser operates unidirectionally. The operation of these three regimes has been experimentally verified with good repeatability. In bidirectional continuous wave operating regime the linewidth is about 110 MHz, whereas in unidirectional operating regime the linewidth is about 40 MHz.

A circular multi quantum well (MQW) InP-based SRL having a radius of 0.6 mm and operating at 1570 nm has been reported in [14]. This laser operates

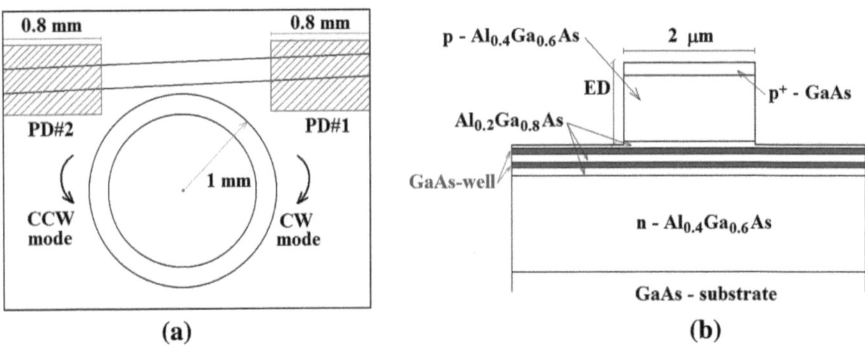

Fig. 5.2 a Architecture of the circular SRL reported in [11] (PD#1 an PD#2 are two photodetectors used to measure optical power of CW and CCW resonant modes). **b** Guiding structure of the SRL reported in [11] (*ED* etch depth)

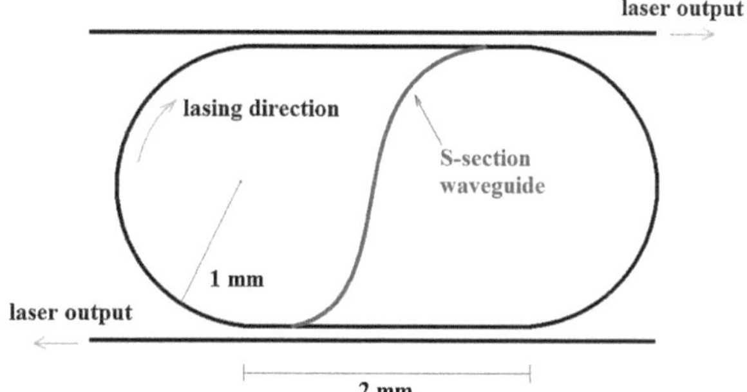

Fig. 5.3 Architecture of the racetrack-shaped SRL reported in [15]

bidirectionally when the injection current is in the range from 125 mA (threshold current) to 135 mA. When injection current ranges from 135 to 220 mA, only one resonant mode (CW or CCW one) is excited. The excited mode switches from one propagation direction to the other as the injection current increases. In this operating condition the system is bistable because if the injection current is increased up to a certain value and a specific lasing direction is selected, the excited resonant mode (either CW or CCW) remains the same also when the injection current is newly reduced and so a change in lasing direction could be expected. When injection current is above 220 mA, the laser is unstable and CW and CCW beams are randomly and alternatively excited.

A large size cavity SRL, with total length >10 mm, operating at 1020 nm has been reported in [15] (see Fig. 5.3). The laser active region is an AlGaAs/GaAs double quantum well (DQW) heterostructure and the resonator is racetrack-shaped. A S-shaped waveguide has been used to connect the opposite straight sections of the ring making the laser intrinsically unidirectional. If the S-section is unbiased, the laser operates bidirectionally only when injection current is just above threshold value (360 mA). If S-section is forward biased, only the CW resonant mode is excited because it is favoured with respect to the CCW mode by the presence of the S-section. A very similar SRL using as active region InAs quantum dots (QDs) embedded inside six 5 nm-thick strained $In_{0.15}Ga_{0.85}As$ quantum wells has been reported in [16]. By this kind of active region a threshold current reduction up to 220 mA has been achieved.

5.1.1.2 Optically Pumped Integrated Ring Lasers

Optically pumped ring lasers are probably less attractive for gyro applications than electrically pumped ones. This is due to the fact that they require a pump laser typically external to the chip and so they appear less suitable than SRLs for the fabrication of a fully integrated optical gyroscope. On the other hand, some optically pumped ring lasers, as those realized in silicon, exhibit very interesting performance in terms of linewidth and power of generated beams and so their use in gyroscopic systems could be attractive.

For the fabrication of optically pumped ring lasers, erbium doped lithium niobate, silicon, III–V semiconductors and π-conjugated polymers have been proposed as system materials.

Optically pumped lasers fabricated in rare-earth doped lithium niobate (LiNbO$_3$) have been investigated since the late 1960s. Doping LiNbO$_3$ crystal with Er^{3+} ions, it is possible to realize high quality optically pumped lasers operating in the wavelength range from 1530 to 1603 nm [17]. Er-doped LiNbO$_3$ can be easily fabricated in the surface layer of a LiNbO$_3$ substrate by diffusion of a thin vacuum-deposited Er layer. Successively a single mode channel waveguide can be realized by the standard titanium diffusion technique. Adopting this technology some distributed feedback and distributed Bragg reflector lasers have been fabricated [18, 19].

The first optically pumped ring laser on Er-doped LiNbO$_3$ has been recently fabricated [20]. This laser, emitting several lines centered around 1603 nm, includes a circular optical resonator having a very large radius (30 mm) and two straight waveguides coupled to the ring (one couples the pump light into the ring and the other one serves as output coupler). Main issues for gyro application of this laser are the quite limited generated optical power (<150 μW) and the fact that single longitudinal mode operation has not been achieved.

An optically pumped ring laser on Er-doped LiNbO$_3$ for gyro application has been patented [21]. The patented architecture includes a racetrack-shaped ring laser in which CW and CCW beams are excited, two electro-optic modulators imposing a phase shift between the optical signals to prevent unidirectional lasing and a U-shaped output waveguide which couples a portion of optical power of counter-propagating waves from the ring laser to the external detection and processing apparatus (see Fig. 5.4).

A silicon large-length (=30 mm) optically pumped ring laser emitting at 1686 nm and based on Raman effect has been recently demonstrated [22] (see Fig. 5.5). The racetrack-shaped laser cavity has been realized by a SOI rib waveguide having a total height of 1.55 μm, etch depth of 0.76 μm and width of 1.5 μm. To reduce optical loss due to free carrier absorption (FCA) and

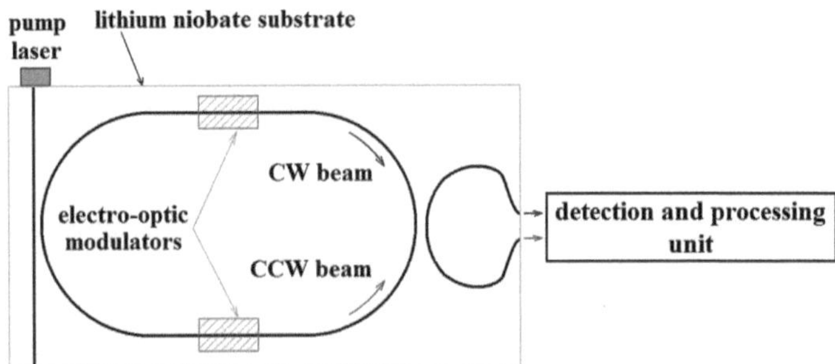

Fig. 5.4 Integrated active optical gyro based on a LiNbO$_3$ optically pumped ring laser [21]

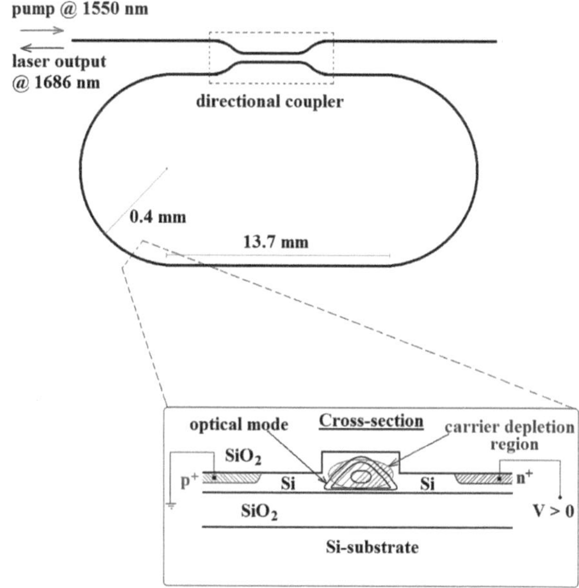

Fig. 5.5 Optically pumped silicon laser based on Raman effect [22]

two-photon absorption (TPA), a p$^+$-doped region and a n$^+$-doped region have been realized in the slab on both sides of the rib structure (see inset in Fig. 5.5). In this manner a p–i–n junction has been created. Reversely biasing the junction (at 25 V), carrier depletion in the region where the optical field is confined occurs, which induces a considerable TPA and FCA decrease and a lasing threshold reduction to 20–40 mW.

Single longitudinal mode operation has been achieved in this laser. Generated beam has a power up to 40 mW, a side mode suppression exceeding 70 dB and a linewidth less than 100 kHz. Although this silicon ring laser requires a pump laser external to the SOI chip providing an optical power of some tens of mW, performance of this optically pumped laser is better than that of large length SRLs mainly in terms of emitted power and linewidth.

The possibility of utilizing silicon ring laser based on Raman effect to sense rotation has been explored in [23] where a silicon ring laser having a total length of 3 cm and emitting at 1549 nm has been designed so that two counter-propagating signals are generated within the laser cavity. If the laser rotates these signals experience the frequency shift predicted by the Sagnac effect. Backscattering due to roughness in waveguide sidewalls induces a dead band within the static characteristic of the sensor ranging from -9.5×10^6 to 9.5×10^6 °/h. To reduce this effect, two thermo-optic modulators, driven by a sinusoidal electrical signal having a frequency of 100 kHz, have been included in the ring cavity. By using this dithering technique, which surely produces an increase of optical loss within the laser cavity, a minimum detectable angular rate value around 300 °/h has been observed. Angle random walk is around 1 °/$\sqrt{h}$.

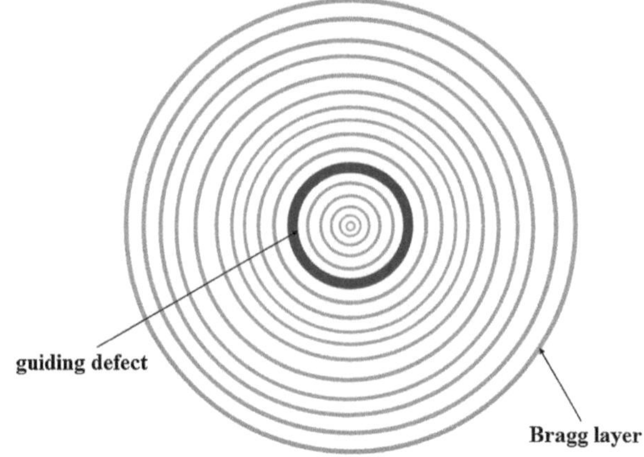

Fig. 5.6 Circular Bragg micro-laser for rotation sensing [25]

An optically pumped ring laser exploiting Bragg reflection instead of total internal reflection as the radial confinement mechanism has been recently proposed [24]. In this device a circular guiding defect, in which optical power is confined, is located within a medium which consists of annular Bragg layers. The lasing structure, schematically shown in Fig. 5.6, has been realized by InGaAsP/InP technology and the active layer includes six quantum wells. The radius of guiding defect is around 5 μm and five internal and ten external Bragg layers have been used to confine generated optical power in the defect. The laser is pumped by optical pulses excitation at 890 nm and emits at 1559 nm (emission direction is perpendicular to the substrate). Threshold power is around 0.7 mW. When this laser rotates, a very interesting dependence of threshold optical power and gain on angular rate has been observed [25]. This effect may enable to sense rotation only by measuring the optical power generated by the circular Bragg micro-laser in the direction perpendicular to the substrate. No data about minimum detectable angular rate are available for this device.

Finally, it is worth to notice that also solid state organic materials for the fabrication of optically pumped ring lasers have been used. A lasing micro-disk with radius of 8 μm on a semiconducting π-conjugated polymer has been reported in [26]. Laser operating wavelength is around 530 nm and it is photoexcited by a Nd:YAG laser producing 100 ps or 10 ns pulses at 532 nm. It is very difficult to envisage the employment of this laser in active integrated optical gyros due to its size and spectral properties of emitted radiation.

5.1.2 *Fully Integrated Active Optical Gyros*

Since the last two decades, design and fabrication of a fully-integrated active optical gyro have been considered a very interesting research target.

The use of a semiconductor laser with a ring type resonator to sense rotation has been patented in 1984 for the first time [27]. The proposed sensor, including the laser and a Y-type coupling element, has been designed to miniaturize and lighten the gyro, reduce power consumption and improve reliability.

A large radius (radius $\sim$1–2 mm) AlGaAs/GaAs SRLs to be monolithically integrated with other readout optical components has been proposed in [28] with the aim of designing a fully integrated active optical angular rate sensor.

A possible readout optical circuit to be integrated on the same chip on which the ring laser is realized has been designed in [29]. This readout circuit (see Fig. 5.7) includes a curved coupler, a MMI coupler and two photodetectors.

A fully integrated active optical gyroscope based on a SRL operating at 845 nm has been proposed, accurately modelled and designed in [30]. In Fig. 5.8 is sketched the active integrated optical gyroscope described in [30], for which an European Patent was granted [31]. Main gyro components are the GaAs/AlGaAs DQW circular SRL, the curved directional coupler for the out-coupling of the two laser-generated beams interfering in the Y-junction, and the photodiode detecting the beating signal resulting from the interference.

The whole optoelectronic sensor is designed to be integrated on a 15×3 mm^2 GaAs substrate. Optical gain is much larger for the quasi-TE mode than for the quasi-TM one. This intrinsic effect of polarization selection inhibits any coupling

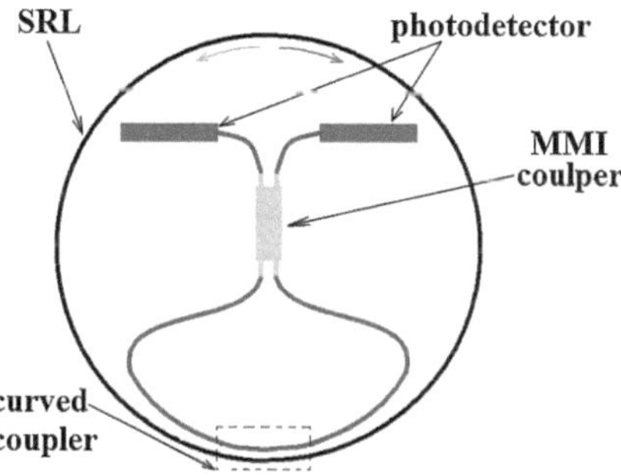

Fig. 5.7 Readout integrated optical circuit proposed for active integrated optical gyroscopes [29]

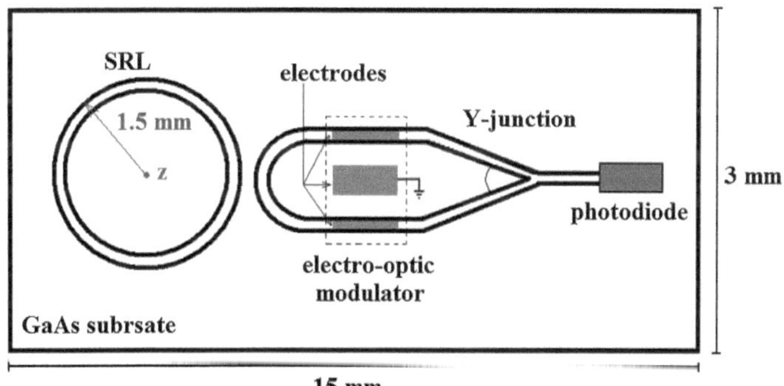

Fig. 5.8 Active integrated optical gyroscope proposed in [30]

between the two polarizations. The gain selectivity obtained by properly designing the gain medium is a crucial aspect of the proposed sensor, since it allows the noise related to the coupling between the two polarizations to be eliminated, with consequent improvement of performance. The guiding structure adopted in the laser includes an $Al_{0.2}Ga_{0.8}As$ barrier which separates the two GaAs quantum wells, two cladding layers to be realized in $Al_{0.2}Ga_{0.8}As$ and a $Al_{0.73}Ga_{0.27}As$ buffer. Waveguide cross-section is shown in Fig. 5.9.

When the sensor rotates the two beams interfering in the Y-junction exhibit different frequencies and the difference between the two frequencies is proportional to the rotation rate. Optical signal at the output of the Y-junction has an amplitude oscillating with a frequency which is equal to the rotation-induced frequency difference. Device angular rate can be estimated by measuring the frequency of electrical signal generated by the photodiode.

In this fully integrated sensor mode locking phenomenon occurs [32]. For reduced angular rate values, the frequency difference between counter-propagating optical waves is equal to zero because of the coupling between the counter-propagating waves induced by waveguide sidewalls roughness. For angular rates less than 210 °/h, in the so called lock-in operating region, the two excited resonant modes exhibit the same frequency so the beating signal generated by the photodetector has only a DC spectral component. In this operating condition (angular rates <210 °/h), the angular rate can be only estimated by measuring the rotation-induced phase shift between counter-propagating optical waves. It depends on the angular rate and on many SRL technological parameters such as radius, backscattering coefficient, scale factor and so on. Then, for reduced angular rate values, the sensor accuracy depends on the accuracy in the estimation of those SRL technological parameters. In lossless condition, minimum detectable angular rate of this integrated gyro has been estimated as around 0.01 °/h.

The electro-optic phase shifter included in the sensor architecture makes more sensitive the measure of rotation-induced phase shift between the laser generated beams. Since the estimation of this phase shift allows to sense rotation also when lock-in occurs, the modulator enhances the sensor resolution in the lock-in

Fig. 5.9 Optical waveguide cross-section of the laser in Fig. 5.8

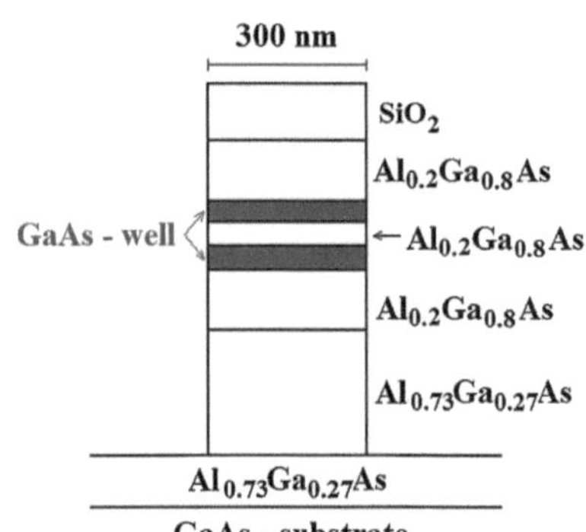

operating region. Moreover rotation sense is detected by this device which imposes a constant phase shift between the optical signals out-coupled from the SRL.

The possibility of detecting the frequency shift due to Sagnac effect only analyzing the spectral components of the voltage signal measured between the terminals of a SRL without extracting the optical power circulating in the laser cavity has been investigated in [33, 34]. According with this approach, a SRL without any other optical component may serve as active integrated optical gyro. To verify this hypothesis an experimental setup consisting of a InGaAsP/InP Fabry-Perot laser, a single mode fiber and the laser driving circuit was used. The laser and the fiber form a 3.6 m long active ring resonator. Keeping constant the laser injection current, the voltage between the laser terminals has been measured through a capacitor, a broadband amplifier and a spectrum analyzer. When a constant rotation rate is applied to the system, a peak at beat frequency appears in the spectrum of the voltage signal measured between the laser terminals. Unfortunately lock-in limited minimum detectable angular rate of this gyroscopic system is around 100 °/h.

An innovative PIC for rotation sensing (see Fig. 5.10) was been proposed in [35, 36]. This architecture was also patented [37]. The device is based on two unidirectionally operating SRLs generating CW and CCW beams which interfere in a Y-junction. Seven photodetectors (electrically isolated by deeply etched isolation trenches) are included in the optoelectronic circuit to monitor device performance.

Ring laser unidirectionality can be achieved by a forward biased (injection current $\sim$60 mA) S-section waveguide connecting the two straight sections of the

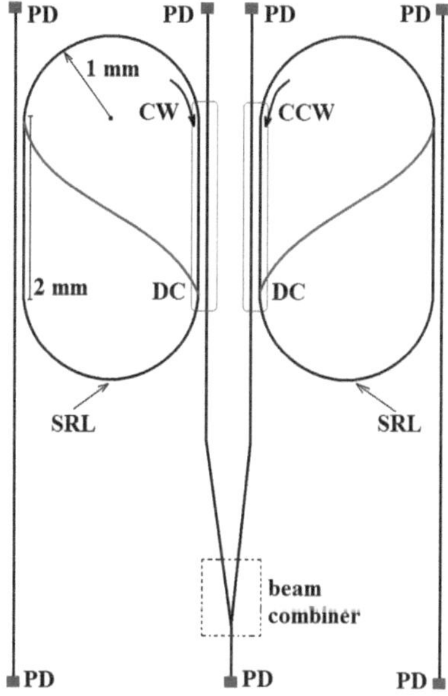

Fig. 5.10 PIC for rotation sensing including two unidirectional SRLs, two directional couplers (DC), seven photodiodes (PDs) and a beam combiner [35]

Table 5.1 PICs proposed for angular rate measurement: performance, geometrical parameters and fabrication steps

Authors	Armenise et al. [30]	Osinski et al. [35]
SRLs included on chip	1	2
Optical components included in the PIC	SRL, U-shaped coupler, electro-optic phase shifter, Y-junction, photodetectors	Twin SRLs, Joule heater, Straight bus waveguides, Y-junction, photodetectors
Operating wavelength (nm)	845	1020 or 1250
Overall PIC size (mm^2)	15 $\times$ 3	6 $\times$ 3
SRL(s) total length (mm)	9.42	10.28
Quantum limit (°/h)	0.01 (theoretical estimation in lossless case)	270–300
Scale factor	1.11 $\times$ 10^5	6–7 $\times$ 10^4
Adopted materials	AlGaAs/GaAs	InGaAs/GaAs/AlGaAs
Active Region	DQW	DQW (or QD)
Fabrication process steps	– (Theoretically proposed device)	MOCVD or MBE, photolithography, ICP etching, BCB deposition, metallization

racetrack shaped cavity. To tune resonant frequencies of the two SRLs, two heaters located along the inner sides of ring ridges have been utilized.

A comparison between the active PICs for gyro applications reported in [30, 35] is in Table 5.1.

Since SRL linewidth usually exceeds 10 MHz, it seems difficult to achieve resolution values less than 100 °/h by active integrated optical gyros. Moreover, attention should to be paid in the design and fabrication process to reduce detrimental effects of the lock-in.

5.2 Passive Integrated Optical Gyros

Typically in passive integrated optical gyroscopes two counter-propagating resonant modes having the same resonance order q are excited. Resonance frequencies of these modes, v_q^{CW} and v_q^{CCW}, are split by rotation. According with Sagnac effect, the difference between v_q^{CW} and v_q^{CCW} is proportional to the cavity angular rate. The operating principle of this device is very similar to that one of the RFOG.

A frequency sensitive passive integrated optical gyro includes an optical cavity, a laser source, some optical devices processing optical signals exciting the resonator and a read-out system consisting of optical and electronic components enabling the estimation of the difference $(v_q^{CW} - v_q^{CCW})$. To perform this

Fig. 5.11 General scheme of
a passive integrated optical
gyroscope

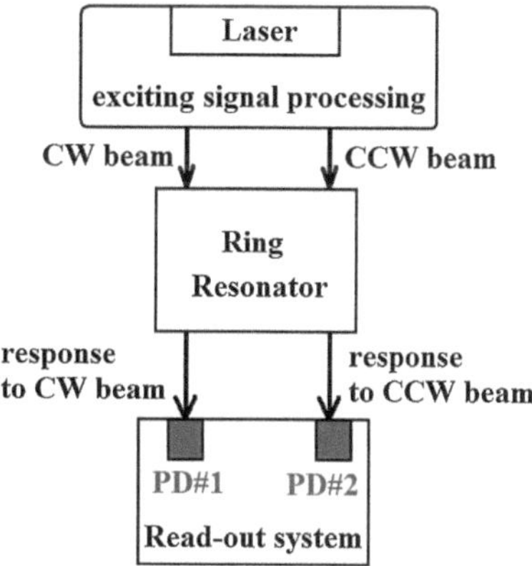

measurement two oppositely-directed laser-generated optical signals have to be coupled into the cavity and resonator response in terms of frequency-dependent reflectivity (R) or transmittivity (T) has to be monitored (R and T exhibit a Lorentzian-like dependence on frequency). Frequency or phase modulation of the two laser-generated beams is usually required. Laser linewidth has to be significantly less than the width of reflectivity or transmittivity peak (typically laser linewidth has to be less than 10 MHz). A general scheme of frequency sensitive passive integrated optical angular rate sensors, proposed for the first time in [38], is shown in Fig. 5.11.

Different kinds of integrated optical cavities have been considered in the design and fabrication of passive gyroscopes, e.g. ring resonators having a circular or a racetrack shape, coupled ring resonators, and photonic crystal resonators.

Read out architectures are based on phase or frequency modulation of signals exciting the optical cavity. Read-out techniques are the same as those already described for the RFOG in Chap. 4. In particular, phase modulation spectroscopy, widely described in previous chapter, is a very effective read-out technique for passive integrated optical gyros. Alternatively frequency modulation spectroscopy can be used. It is based on the sinusoidal modulation of the laser generated beams which are coupled within an integrated resonator [39]. The implementation of this read-out technique requires the use of a frequency modulated laser which is a very critical and expensive device.

5.2.1 Passive Integrated Optical Gyros Based on a Ring Resonator

In the last decade, integrated optical ring resonators have been largely investigated, and a number of application fields have been proposed for these components, which have been fabricated using a wide range of materials, i.e. glass, silica, III–V semiconductors, silicon, silicon oxynitride, polymers, and lithium niobate [40–43].

One of the most interesting applications of large dimensions (in the range of millimetres) ring resonators is surely the one related to the integrated passive gyros.

A ring resonator usually includes a curved optical waveguide having a ring shape and one or two straight bus waveguides. The bus waveguides and the ring are evanescently coupled. For some applications as chemical sensing, wavelength filtering and multiplexing/demultiplexing, only one input beam is launched in a bus waveguide to excite the resonator. When the ring is used for rotation sensing two input signals are launched, simultaneously or not, in the bus waveguides to excite the cavity for both CW and CCW propagation directions. If a two bus waveguide architecture is exploited, the two input beams can be launched in two different bus waveguides (see Fig. 5.12a) or in the two opposite ends of the same bus waveguide (see Fig. 5.12b). Consequently there are two possible configurations for the ring excitation and resonance frequency measurement. In the first case output ports are called drop ports, in the second configurations are the through ports.

Using a one bus waveguide architecture (see Fig. 5.12c), each end of the bus can be utilized either as input or output port. In this case, two circulators or two switches at both ends of the bus have to be used to excite the resonator in CW and CCW direction and to monitor spectral response at the through port. As previously pointed out, the excitation of the cavity in the two opposite directions may be simultaneous or not.

The two beams exciting the resonator must have the same power with a very reduced tolerance. A difference in the power of the two signals coupled within the

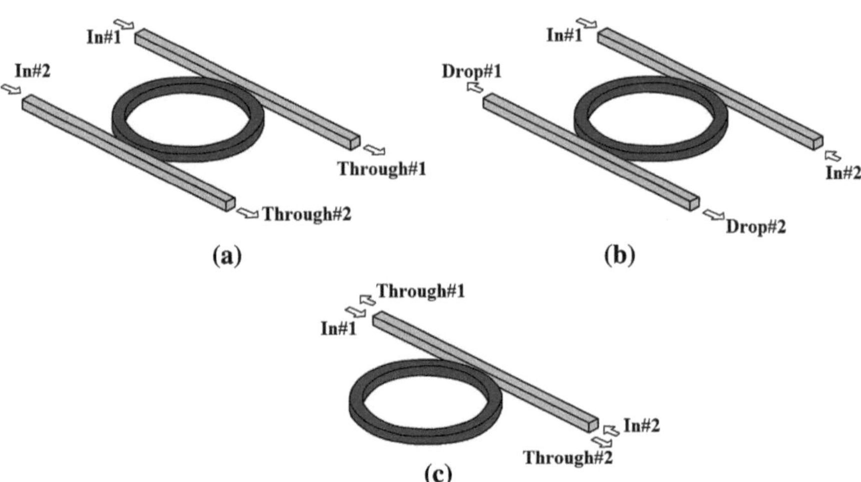

Fig. 5.12 a, b Ring resonator coupled with two bus waveguides. Input ports on two different waveguides (a) or on the same waveguide (b). Ring resonator coupled with one bus waveguide (c)

cavity induces a bias in the sensor output. Since this power difference is unpredictable, the bias value is unpredictable too.

The physical origin of this bias is a Kerr-like nonlinearity of materials of the ring resonator waveguide. The refractive index n of all materials exhibiting this Kerr-like nonlinearity, e.g. Silica, Si, InP, can be written as:

$$n = n_0 \sqrt{1 + \bar{n}_2 \frac{\left|\vec{E}\right|^2}{Z_0}} \tag{5.3}$$

where n_0 is the linear part of the refractive index, $\bar{n}_2$ is the nonlinear-index coefficient expressed in m^2/W, Z_0 is the free space impedance and $\vec{E}$ is the electric field vector.

The effective index n_{eff} of a waveguide whose core is realized by a material exhibiting a Kerr-like nonlinearity can be expressed as:

$$n_{eff} = n_{eff,0} + \bar{n}_{2,core} \frac{P_m}{A_{eff}} \tag{5.4}$$

where P_m is optical power carried out by the propagating mode, $n_{eff,0}$ is the linear contribution to the effective index, $\bar{n}_{2,core}$ is the nonlinear-index coefficient of the material forming the waveguide core, and A_{eff} is the propagating mode effective area that can be approximated by the core area.

Tacking into account the Kerr effect and assuming the cavity at rest, the two resonance frequencies for CW and CCW direction are given by:

$$v_q^{CW} = \frac{qc}{p\left(n_{eff,0} + \bar{n}_{2,core}\frac{P_{CW}}{A_{eff}}\right)} \qquad v_q^{CCW} = \frac{qc}{p\left(n_{eff,0} + \bar{n}_{2,core}\frac{P_{CCW}}{A_{eff}}\right)} \tag{5.5}$$

where q is the order of the considered resonant mode, P_{CW} and P_{CCW} are the optical powers of the two modes circulating in the cavity. We have assumed a not simultaneous excitation.

Kerr effect induced bias is equal to:

$$\Delta\Omega_{Kerr} = S^{-1}\left(v_q^{CW} - v_q^{CCW}\right) = \frac{4 A_{eff} n_{eff,0} \bar{n}_{2,core}\, \delta P\, v_0\, S^{-1}}{4\left(A_{eff} n_{eff,0} + \bar{P}\,\bar{n}_{2,core}\right)^2 + \left(\bar{n}_{2,core}\,\delta P\right)^2} \tag{5.6}$$

where S is the gyro scale factor, δP is the difference between P_{CW} and P_{CCW}, v_0 is the gyro operating frequency and $\bar{P}$ is the average value between P_{CW} and P_{CCW}. Since typical $\bar{n}_{2,core}$ values are less than 10^{-16} W/m^2, we obtain $\bar{P}\,\bar{n}_{2,core} \ll A_{eff} n_{eff,0}$ and $\left(\bar{n}_{2,core}\,\delta P\right)^2 \ll A_{eff} n_{eff,0}$. Then an approximated expression for Kerr effect induced bias can be written:

$$\Delta\Omega_{Kerr} \cong \frac{\bar{n}_{2,core}}{S A_{eff} n_{eff,0}} \delta P\, v_0 \times \frac{3600 \times 180}{\pi}\ (^{\circ}/\text{h}) \tag{5.7}$$

Equation 5.7 clearly shows that the bias drift induced by the Kerr effect is proportional to the unpredictable difference between optical powers of the CW and CCW signals propagating in the resonant cavity.

Resonator spectral responses at the through ports have Lorentzian-like shape with a minimum at the two resonance frequencies for CW and CCW propagation directions. The spectral responses at the drop ports exhibit a maximum at the two resonances and they have a Lorentzian-like shape, too. In all cases, resonator spectral response is periodical and the response period (i.e. free spectral range, FSR) is inversely proportional to the ring radius.

Ring resonator spectral responses at the through port in case of one or two I/O bus waveguides are given by:

$$T_{1busWG}(\lambda) = \left| \frac{\sqrt{\gamma - \kappa^2} - \gamma\, e^{i\beta p} e^{-\alpha p/2}}{1 - \sqrt{\gamma - \kappa^2}\, e^{i\beta p} e^{-\alpha p/2}} \right|^2$$

$$T_{2busWGs}(\lambda) = (\gamma - \kappa^2) \left| \frac{1 - \gamma\, e^{i\beta p} e^{-\alpha p/2}}{1 - (\gamma - \kappa^2)\, e^{i\beta p} e^{-\alpha p/2}} \right|^2 \tag{5.8}$$

where α is the attenuation coefficient due to propagation loss within the ring, κ^2 is the optical power fraction passing from the bus waveguide to the ring, γ (≤ 1) is the coefficient taking into account the loss due to bus-ring coupling, β is the propagation constant of optical modes within the ring and L is the resonator length. $T(\lambda)$ reaches its minimum value when $\beta L = 2q\pi$, maximum for $\beta L = 2\,(q + 1)\pi$, being q a natural number known as resonance order.

Starting from the spectral response at through port, resonator finesse F and quality factor Q can be defined as:

$$F = \frac{FSR}{\delta\lambda} \quad Q = \frac{\lambda_{q,0}}{\delta\lambda} \tag{5.9}$$

where $\delta\lambda$ is the resonance spectral width, namely full wave half maximum, FWHM, and $\lambda_{q,0}$ is the resonance wavelength.

Performance of passive integrated gyros based on a ring resonator strongly depends on optical cavity size and quality factor. Independently of the readout system choice, shot noise-limited minimum detectable angular rate can be written as:

$$\delta\Omega = \frac{c\sqrt{2}}{SQ\lambda_0}\Pi \times \left(\frac{3600 \times 180}{\pi} \right) \left({}^{\circ}/\sqrt{h} \right) \tag{5.10}$$

where λ_0 is the sensor operating wavelength.

Shot noise induced ARW can be written as:

$$W_{ARW} = \frac{c\sqrt{2}}{SQ\lambda_0\sqrt{B}}\Pi \times \left(\frac{60 \times 180}{\pi} \right) \left({}^{\circ}/\sqrt{h} \right) \tag{5.11}$$

Dimensions of the integrated passive resonator influence gyro scale factor and cavity quality factor depends on loss and resonator length. To achieve $\delta\Omega < 5$ °/h and ARW < 0.02 $°/\sqrt{h}$, a quality factor around 10^6 and an resonator length in the range of centimetres are at least required.

Technologies enabling the realization of low-loss optical waveguides are usually employed in the integrated optical cavity realization because loss suffered by the resonant mode strongly influences the cavity quality factor.

Silica-on-silicon technology allows very low-loss (<0.1 dB/cm) optical waveguides operating at 1.55 μm. Typically these waveguides are fabricated by flame hydrolysis deposition and reactive ion etching. On a silicon substrate, two SiO_2 layers are deposited by flame hydrolysis. The second layer, which is Ge doped, has a higher refractive index with respect to the first one. After the deposition, the two layers are consolidated in an oven at 1300°C and then the guiding layer is partially removed by photolithography followed by reactive ion etching. Finally a silica cladding is deposited by flame hydrolysis and then consolidated. Propagation loss of these waveguides depends on index contrast Δ between the guiding and the cladding layer and propagation loss around 0.02–0.03 dB/cm has been achieved for $\Delta < 1\%$. Bending loss suffered by these waveguides exponentially decreases with curvature radius. To achieve negligible bending loss, a curvature radius larger than a few millimetres is required.

Some ring resonators having a very large quality factor have been fabricated in silica-on-silicon technology. The largest experimentally measured quality factor, 2.4×10^7, was obtained by a resonator operating at 1.55 μm and employing a waveguide exhibiting a 5 μm × 5 μm squared core region realized depositing a phosphorus doped SiO_2 core layer and a boron and phosphorus doped glass top cladding layer on thermally grown silicon oxide [44]. Waveguide index contrast was 0.7% and estimated propagation loss around 0.024 dB/cm.

To further enhance quality factor of resonators in silica-on-silicon technology, the hybrid integration of two semiconductor optical amplifiers (SOAs) within a silica-on-silicon ring resonator was proposed (see Fig. 5.13) [45–48]. The two SOAs compensate optical loss suffered by resonant mode and so a very high

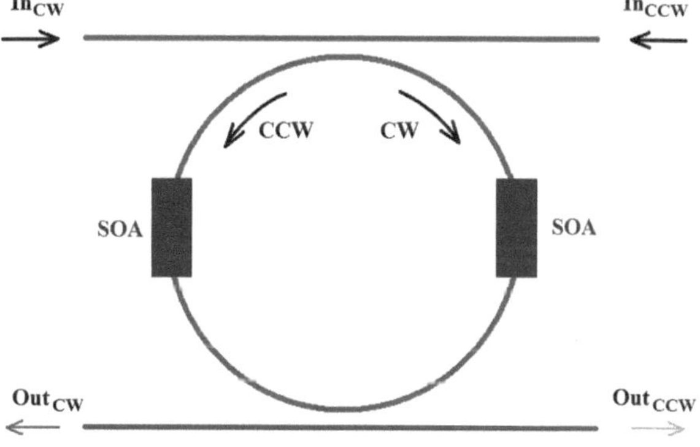

Fig. 5.13 Ring resonator in which propagation loss is compensated with two SOAs [45]

quality factor is expected [49, 50]. For a ring radius of 10 mm and a coupling efficiency between the straight waveguides and the ring resonator equal to 0.1%, a quality factor as high as 2.9×10^8 was calculated, neglecting the effect of spontaneous emission noise. Taking into account, also, the effect of amplified spontaneous emission noise, a minimum detectable rotation rate around 10 °/h has been predicted if this resonator is used as building-block in a passive integrated optical gyroscope [51].

The SOAs, realized by the InP-based technology, is not monotonically integrated on the silica chip and this may produce some problems in the power transfer between the ring and the amplifiers and could generate back-reflections at the interfaces between the silica-on-silicon ring and the SOAs.

The first PIC properly designed to sense rotation is reported in [52] (see Fig. 5.14). The PIC, including a 14.8 cm long integrated ring resonator was realized using silica-on-silicon technology to achieve low propagation loss and an integrated binary phase shift keying (BPSK) modulator based on thermo-optic effect was included in the sensor to limit the power transfer between CW and CCW beams. Angular velocity was measured by alternatively locking laser frequency to CW and CCW resonance frequencies through four integrated switches based on Mach-Zehnder interferometer.

In this sensor, the laser source, the photodetectors and read-out electronic circuitry are outside the chip. The gyro operates at 1550 nm with $\delta\Omega \sim 10$ °/h.

More recently another PIC for rotation sensing including only a ring resonator having a radius of 9.5 mm (total length around 6 cm), three couplers and only one bus waveguide has been proposed using the silica-on-silicon technology [53].

Fig. 5.14 Frequency-sensitive passive integrated optical gyro in silica-on-silicon technology [52]

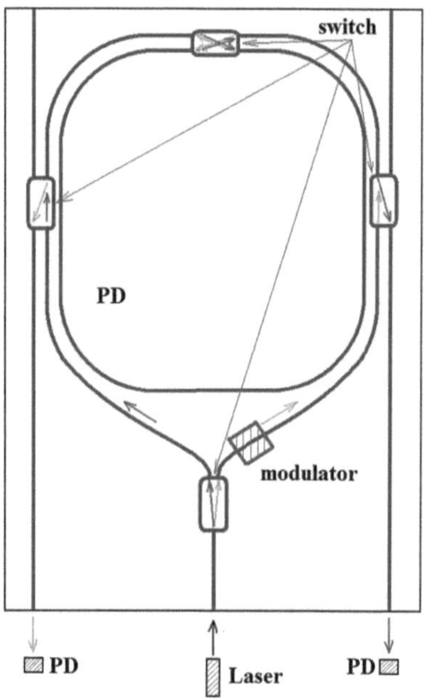

Employed guiding structure exhibits very reduced loss and so the resonator quality factor is quite large (3.12×10^6). Sensor minimum detectable angular rate is around 15 °/h. All the read-out system including a fiber laser, two phase modulators, two acousto-optic frequency shifters, two polarization controllers, two photodetectors and two LIAs is out of chip. To enhance the gyroscope sensitivity, the same research group has recently proposed to use a multi-turn ring resonator in silica-on-silicon technology having a total length until 77 cm [54]. The optical cavity is formed by a number of crossed waveguides that have been in deep investigated demonstrating that optimum intersection angle is close 90°. A minimum detectable angular rate in the range 0.1–1 °/h has been theoretically predicted for a passive gyro including this multi-turn resonator.

It is well known that other two technologies for fabricating low-loss waveguides are ion exchange in glass and titanium in-diffusion in lithium niobate ($LiNbO_3$). Minimum value of propagation loss in $LiNbO_3$ waveguides has been demonstrated to be as low as 0.03 dB/cm at 1.55 μm [55], whereas ion exchange in glass substrate enables to achieve loss less than 0.1 dB/cm [56]. Although these loss values are rather low, they are quite larger than those exhibited by silica-on-silicon waveguides. On the other hand, the technological process to fabricate ion-exchanged glass waveguides is relatively simple and economical and a variety of active dopant species may be introduced into the glass enabling the compensation of propagation loss and the realization of optically pumped lasers. $LiNbO_3$ technology attractiveness for the realization of passive gyroscopes based on ring resonators is eminently due to the possibility to monolithically integrate the ring with an optically pumped laser ($LiNbO_3$ can be doped with erbium) and electro-optic modulators.

Silver ion exchange in a commercially available silicate glass (Schott IOG-10) allowed to fabricate a circular ring resonator operating at 1550 nm and having a radius equal to 8 mm to be employed in a passive integrated optical gyro [57]. E-beam evaporation was used to deposit a 150 nm thick layer of titanium onto the glass substrate, conventional photolithography step followed to pattern the titanium with 2 μm wide channel openings. Ion exchange performed in a mixed melt of silver nitrate and sodium nitrate, and the final was the thermal annealing. Measured propagation loss within the resonator is around 0.1 dB/cm and quality factor has been calculated as equal to 2×10^6. On the same substrate on which the resonator has been fabricated, two straight bus waveguides to couple CW and CCW beams within the resonator and a Y-splitter have been realized.

A PIC for rotation sensing fabricated by sputtering deposition of glass and reactive ion etching was recently reported [58]. The system includes a ring resonator and a tuneable optically pumped laser realized by doping glass with erbium. This is not a fully integrated passive optical gyroscope because the pump laser and the two photodetectors of the read-out system are outside the chip. No data on this gyro performance are available.

Propagation loss compensation by the optical gain induced by a pump signal has been proposed to achieve Q value exceeding 10^7 in glass ring resonators. In [59] a racetrack-shaped compensated resonator on neodymium doped-silicate glass has been reported. The cavity operates at 1.02 μm and has a total length of 56.07 mm. The pump signal, having a power of 150 mW, is provided by a semiconductor laser emitting at 0.83 μm. The device architecture includes two bus waveguides, one for the signal (used to excite the resonator and to measure resonator spectral response) and the other for the pump (see Fig. 5.15). The compensation of propagation loss within the ring allows to obtain very large finesse and quality factor ($F = 250$, $Q = 1.89 \times 10^7$).

A high-Q LiNbO$_3$ integrated optical ring resonator designed for angular rate sensing is reported in [60]. The resonator, which exhibits a radius equal to 30 mm and a quality factor of 2.4×10^6, has been fabricated in a Z-cut LiNbO$_3$ substrate by thermal in-diffusion (1060°C) of 7 μm wide and 100 nm thick Ti-stripes. One bus waveguide architecture has been used. Optical loss within the resonator is about 0.03 dB/cm.

The use of this ring resonator for rotation sensing has been both theoretically and experimentally investigated. Theoretical minimum detectable angular rate is around 7 °/h but drift in laser frequency dramatically degrades the resolution. Preliminary experimental result obtained by the authors was $\delta\Omega = 36{,}000$ °/h.

III–V semiconductors have the potential of enabling the fully monolithic integration of the passive integrated gyro. Unfortunately, passive ring resonators fabricated in III–V semiconductors usually exhibit a reduced quality factor and dimensions in the range of tens of microns [61].

Two approaches have been explored to enhance quality factor of ring resonators fabricated by III–V semiconductors. First approach involves the compensation of propagation loss of the resonant mode by inserting one or two SOAs in the ring. The second one is the use of a large radius ring and a low index contrast waveguide.

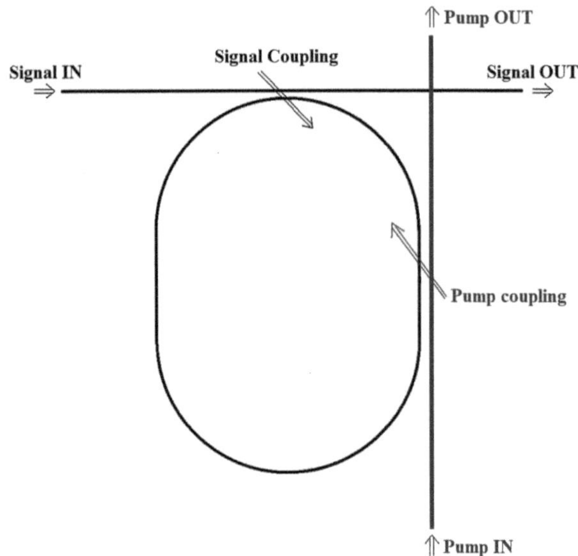

Fig. 5.15 Compensated ring resonator realized by ion exchange in glass [59]

A SOA-compensated InP ring resonator has been reported in [62]. In this device, a SOA is included within the ring according to the architecture shown in Fig. 5.16. The SOA waveguide and the passive waveguide have different geometric configurations and are butt coupled. At the beginning of the fabrication process the passive ring is realized, then a section of the passive ring is removed and the SOA is realized on the InP substrate by selective re-growth. The total length of the ring is about 7 mm and the quality factor is 2.2×10^5. The same quality factor value has been reported for a InP ring resonator including two SOAs realized by another research group [63]. Main issues of this approach to enhance the quality factor of InP ring resonators are related to the technological solution enabling to integrate a passive waveguide and a SOA on the same InP substrate. Moreover SOA integration in a passive ring can produce an increase of back-reflections within the resonator and introduce amplified spontaneous emission noise in the cavity.

Decreasing the index contrast between the guiding layer and the cladding material, an InGaAsP/InP passive waveguide exhibiting relatively low propagation loss (<1 dB/cm) can be realized. In [64], a fully buried InGaAsP/InP waveguide having $\Delta = 5.74\%$ has been proposed for the realization of a high-Q uncompensated InP ring resonator. Propagation loss in this guiding structure (cross-section in Fig. 5.17) has been experimentally proved to be 0.8–0.9 dB/cm. A ring resonator having a radius of 0.2 mm and a quality factor equal to 1.3×10^5 has been fabricated by using this waveguide. In [65] was demonstrated that an InGaAsP/InP waveguide exhibiting propagation loss less than 0.5 dB/cm require Δ less than 5%. A large-radius (>1 mm) ring resonator made by this waveguide has a quality factor $>10^6$. Minimum detectable angular rate of a passive gyro including a InP ring resonator having a total length of 20 cm and a quality factor around 1.5×10^6 may be less than 10 °/h [66].

Maximum achieved quality factor in SOI ring resonators is around 1.5×10^5. This limitation in quality factor makes very difficult to realize a passive integrated optical gyroscope having $\delta\Omega < 10$ °/h by using a SOI ring resonator.

Table 5.2 summarizes performance of high-Q ring resonators reported in literature. Minimum detectable angular rate achievable if they are used as sensing element of a gyro has been estimated. From reported data, it is clear that the typical value of $\delta\Omega$ for a passive integrated optical gyroscope based on a ring

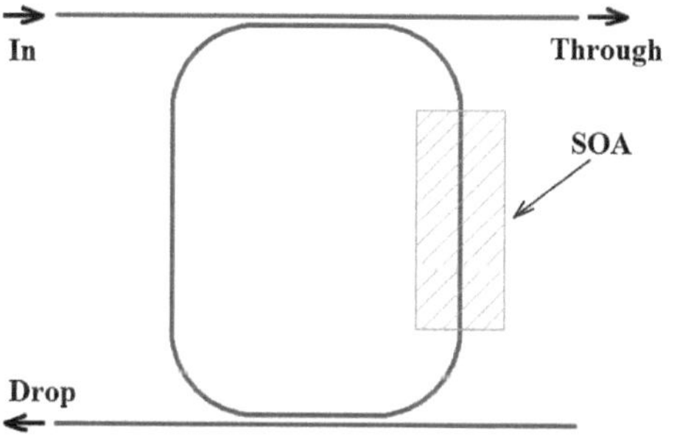

Fig. 5.16 SOA compensated optical resonator architecture

Fig. 5.17 Cross-section of the low loss InGaAsP/InP waveguide minimizing the scattering loss

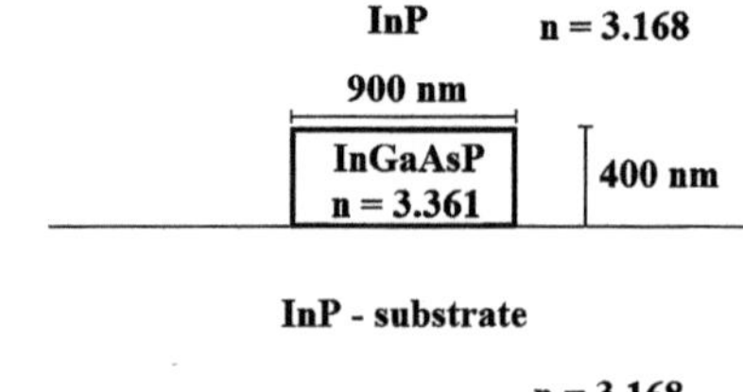

Table 5.2 Main performance parameters of high-Q ring resonators reported in literature

Authors	Cavity length (mm)	Scale factor (Hz/rad s^{-1})	Quality factor	Estimated $\delta\Omega$ (°/h)	Estimated ARW (°/$\sqrt{h}$)	Technology
Adar et al. [44]	188.4	3.87×10^4	2.3×10^7	1	0.0037	Silica-on-silicon
Suzuki et al. [52]	148	–	–	10	0.037	Silica-on-silicon
Ciminelli et al. [45]	94.8	1.90×10^4	2.9×10^8 (theoretical result)	0.2	7.5×10^{-4}	Silica-on-silicon (SOA comp.)
Ma et al. [53]	59.7	1.20×10^4	3.1×10^6	15	0.056	Silica-on-silicon
Li et al. [57]	50.2	1.03×10^4	2×10^6	50	0.19	Ag$^+$ ion exchange in glass
Hsiao et al. [59]	56.1	1.66×10^4	1.9×10^7	3	0.01	Ion exchange in N$_2$O$_3$ doped glass
Vannahme et al. [60]	188.4	3.87×10^4	2.4×10^6	7	0.026	Ti:LiNbO$_3$

resonator is in the range 1–10 °/h and the typical value of ARW is in the range 4×10^{-3}–4×10^{-2} °/$\sqrt{h}$.

As already mentioned, the bias drift in passive integrated optical gyroscopes is induced by Kerr effect. It depends on the difference δP between optical powers of the CW and CCW signals propagating in the resonant cavity. Considering a rotation sensor including a silica-on-silicon ring resonator having $S = 2 \times 10^4$ Hz/rad/s, $A_{eff} = 25$ μm^2, $n_{eff,0} = 1.5$, $v_0 = 193.5$ THz, and $\bar{n}_{2,core} = 2.6 \times 10^{-20}$ m^2/W (typical value for SiO$_2$) the dependence of gyro bias on δP has been plotted in Fig. 5.18 by using Eq. 5.7. To realize a silica-on silicon gyro with bias drift <0.5 °/h it is necessary that δP is less than 0.3 μW. For $\delta P < 0.1$ μW we have $\Delta\Omega_{Kerr} < 0.15$ °/h. For silica-on-silicon passive gyros

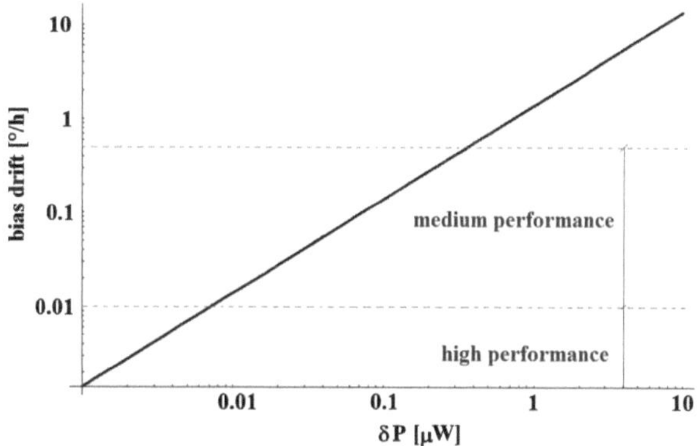

Fig. 5.18 Bias instability induced by δP (log–log sale)

designed to assure high performance (bias drift $< 0.01°/\text{h}$), the value of δP must be accurately controlled (it has to be in the range of a few nanowatts).

5.2.2 Passive Integrated Optical Gyros Based on Coupled Ring Resonators

Integrated optical ring resonators have been widely studied to slow light by decreasing its group velocity [67]. Two structures have been considered to this purpose (see Fig. 5.19): the Coupled-Resonator Optical Waveguide (CROW) [68] and the Side-Coupled Integrated Spaced-Sequence of Resonators (SCISSOR) [69, 70]. CROW consists of a chain of coupled resonators in which light propagates because of the coupling between adjacent resonators. SCISSOR consists of sequence of resonators evanescently coupled with a bus waveguide. The resonators

Fig. 5.19 a CROW architecture. **b** SCISSORs architecture

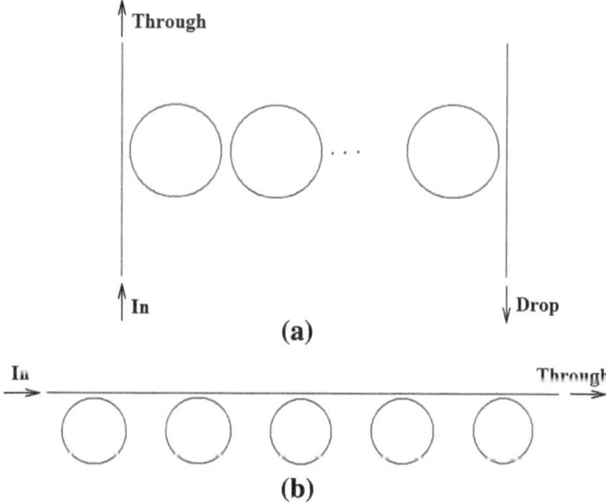

are sufficiently close to the bus waveguide to achieve evanescent coupling but they are spaced far enough from one another so that resonator-to-resonator coupling can be considered as negligible.

If two optical beams counter-propagate in a dispersive resonant structure, they experience a Sagnac phase shift. Rotation induced phase shift is proportional to the group index of the dispersive structure. Then the Sagnac phase shift would be enhanced when the two beams counter-propagate in a closed loop realized by a CROW or a SCISSOR where the group index of light may be quite large [71–73].

A CROW-based passive integrated optical gyro was theoretically proposed and modelled in [74] (see Fig. 5.20). A 3 dB power divider based on a directional coupler splits the laser beam into two signals which are launched in the two ends of the CROW wrapped around itself. If the gyro is motionless, the two signals acquire an identical phase shift during the propagation in the CROW. When the gyro rotates, this phase shift is different for the two signals propagating in CW and CCW direction so they are no longer in phase after the propagation in the CROW. Signals at the output of the CROW has a rotation induced phase shift $\Delta\varphi$ and interfere in the directional coupler. The power of optical signals at the output of the two coupler ports depends on $\Delta\varphi$ according with following equations:

$$P_{out,1} = P_{in}\ \cos^2\left(\frac{\Delta\varphi}{2}\right) \quad P_{out,2} = P_{in}\ \sin^2\left(\frac{\Delta\varphi}{2}\right) \tag{5.12}$$

where P_{in} is the optical power of the laser beam.

Fig. 5.20 CROW-based integrated gyro architecture [74]

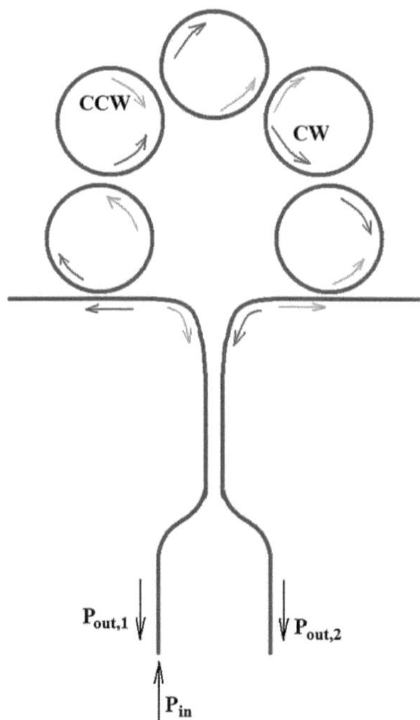

Measuring the power at the two outputs of the coupler, it is possible to estimate $\Delta\varphi$ and the rotation rate. In a CROW including nine rings having a radius of 25 μm, an angular rate of 1°/h generates a change around 4% in $P_{out,2}$. Unfortunately, to achieve this performance, the quality factor of each ring resonator has to be around 10^7. Such a high quality factor has been achieved only by large radius silica-on-silicon ring resonators and not by micro-rings realized, for example, using the SOI technology. A decrease in the quality factor of rings down to 10^4–10^5, significantly degrades the rotation sensor sensitivity.

A similar integrated optic rotation sensor has been proposed in [75]. Also in this device the phase shift between two counter propagating beams is measured to sense rotation. In this device, the integrated passive gyroscope includes a directional coupler, a U-shaped curved waveguide and an optical resonator including two coupled rings having a different radius. If the sensor is rotating, the beams out-coupled from the optical cavity exhibit a phase shift $\Delta\varphi$ equal to:

$$\Delta\varphi = \frac{n_g}{n_{eff}} \times \frac{8\pi^2 R_1^2}{c\lambda}\Omega \tag{5.13}$$

where n_g is the group index of light in the ring-in-ring resonator and n_{eff} is the effective index of the optical signals propagating in the sensor, λ is the device operating wavelength and R_1 is the radius of the largest ring.

In this device, the gyro scale factor is the product between the typical scale factor of a phase sensitive gyroscope (e.g. IFOG) and the ratio n_g/n_{eff}, where:

$$\frac{n_g}{n_{eff}} = \frac{\partial T_\phi}{\partial \phi_1} \tag{5.14}$$

with

$$\phi_1 = \beta_1(2\pi R_1), \tag{5.15}$$

β_1 is the propagation constant within the largest ring, and T_ϕ the wavelength-dependent phase response of the resonator. It results:

$$T_\phi = \arg\left[\frac{E_{in}}{E_{out}}\right] \tag{5.16}$$

where E_{in} and E_{out} are the complex amplitudes of signals at the two port of the cavity. The T_ϕ dependence on ϕ_1 exhibits the maximum slope in correspondence of the resonance wavelength of the largest ring. Then the ratio n_g/n_{eff} can be maximized if the sensor operating wavelength corresponds exactly to this resonance wavelength.

For the sensor proposed in [75] having $R_1 = 16.4$ cm the ratio n_g/n_{eff} is around 50 at $\lambda = 1.55$ μm. A rotation of 10 °/h induces $\Delta\varphi = 4.6 \times 10^{-7}$ rad.

A passive integrated gyro based on a SCISSOR structure has been theoretically proposed in [76] (see Fig. 5.21). The sensor includes a circular waveguide coupled to a quite large number of high-Q micro-resonators. The two beams

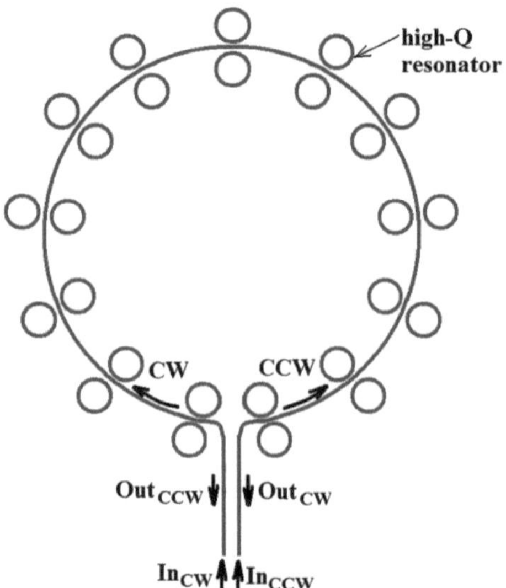

Fig. 5.21 Passive integrated optical gyro including a circular waveguide coupled to high-Q micro-resonators [76]

counter-propagating in the circular guiding structure acquire a rotation-induced phase shift. This phase shift is enhanced by the low group velocity of light in its circular path within the sensor. No data on the device performance have been provided in [76].

Effect of rotation on spectral response at the drop port of a CROW including a quite large number of coupled integrated optical resonators has been theoretically investigated in [77]. Spectral response at the drop port of a motionless CROW exhibits a periodical sequence of transmission and forbidden bands. Rotation induces additional forbidden bands in the CROW response. The width and the dip of these additional forbidden bands are directly proportional to CROW angular rate. The scale factor relating the width of rotation-induced forbidden bands and the angular rate (in rad/s) is equal to 67.5 for a CROW including 29 ring resonators having a radius equal to 25 μm [77]. To observe the presence of the rotation-induced forbidden band a very large rotation rate is required. For example in the device including 29 rings, a rotation rate around 10^8 rad/s ($=2 \times 10^{13}$ °/h) is required to begin to observe the presence of the rotation-induced forbidden band in the CROW spectral response at the drop port.

5.2.3 Passive Integrated Optical Gyros Based on a Photonic Crystal Cavity

Photonic crystals (PhCs) are ordered structures in which two media with different refractive index are arranged in a periodic form with a periodicity on the order of hundreds of nanometers or less. These sub-micrometer structures can be designed so that frequency bands in which the optical propagation is inhibited for all propagation directions are induced [78]. Depending on spatial periodicity of the

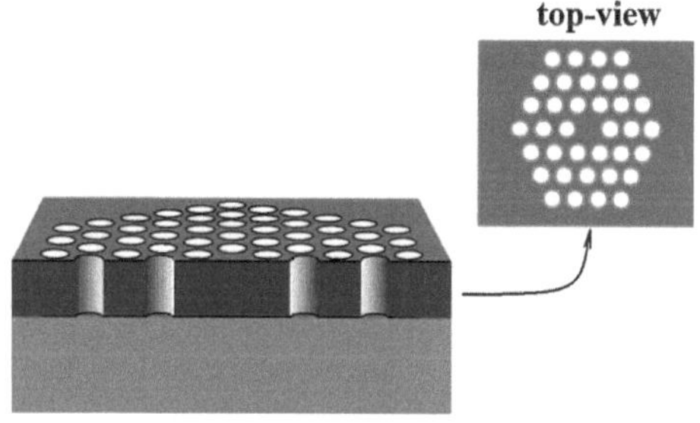

Fig. 5.22 Structure of a high-Q PhC micro-cavity

structure, we distinguish one-dimensional (1D), bi-dimensional (2D) and three-dimensional (3D) PhCs.

A very attractive structure for PhCs fabrication is a thin semiconductor slab waveguide perforated with a periodic distribution of a 2D lattice of air holes. In this structure light confinement in vertical direction is obtained by total internal reflection and in-plane light confinement is due to 2D lattice of holes [79]. Introducing a single point defect in this structure (for example by modifying radius of one hole or eliminating one hole), a high-Q optical micro-cavity can be obtained (see Fig. 5.22).

Very recently, B. Z. Steinberg and A. Boag predicted that in a PhC micro-cavity supporting two degenerate resonance modes a symmetrical splitting of the degenerate resonance frequency $v_{q,0}$ into two distinct resonance frequencies $v_{q,0} \pm \Delta v/2$ is induced by rotation [80]. Difference Δv is proportional to the angular velocity of the PhC cavity. Δv has been numerically estimated for a PhC micro-cavity having a quality factor exceeding 10^5. In this case, the scale factor relating Δv and angular rate Ω (in rad/s) is around 1.6×10^{-2}. This very reduced value of the scale factor imposes further investigation for PhC micro-cavities practical employment to sense rotation.

The employment of coupled PhC micro-cavities has been discussed in [81]. Using 12 coupled micro-cavities a scale factor less than 3 has been obtained. This value is very far from typical scale factor values of passive integrated optical gyros.

References

1. Nagarajan, R., et al.: Large-scale photonic integrated circuits. IEEE J. Sel. Top. Quantum Electron. **11**, 50–65 (2005)
2. Liao, S., Wang, S.: Semiconductor injection lasers with a circular resonator. Appl. Phys. Lett. **36**, 801–803 (1980)
3. Jezierski, A.F., Laybourn, P.J.R.: Integrated semiconductor ring lasers. IEE Proc. **135**, 17–24 (1988)
4. Krauss, T.F., Laybourn, P.J.R., Roberts, J.S.: CW operation of semiconductor ring lasers. Electron. Lett. **26**, 2095–2097 (1990)
5. Hohimer, J.P., Craft, D.C., Hadley, G.R., Vawter, G.A., Warren, M.E.: Single-frequency continuous-wave operation of ring diode lasers. Appl. Phys. Lett. **59**, 3360–3362 (1991)

6. Choi, S.J., Djordjev, K., Choi, S.J., Dapkus, P.D.: Microdisk lasers vertically coupled to output waveguides. IEEE Photonics Technol. Lett. **15**, 1330–1332 (2003)
7. Zhang, R., Ren, Z., Yu, S.: Fabrication of InGaAsP double shallow ridge rectangular ring laser with total internal reflection mirror by cascade etching technique. IEEE Photonics Technol. Lett. **19**, 1714–1716 (2007)
8. Chen, Q., Hu, Y.-H., Huang, Y.-Z., Du, Y., Fan, Z.-C.: Equilateral-triangle-resonator injection lasers with directional emission. IEEE J. Quantum Electron. **43**, 440–444 (2007)
9. Van Campenhout, J., Rojo-Romeo, P., Regreny, P., Seassal, C., Van Thourhout, D., Verstuyft, S., Di Cioccio, L., Fedeli, J.-M., Lagahe, C., Baets, R.: Electrically pumped InP-based microdisk lasers integrated with a nanophotonic silicon-on-insulator waveguide circuit. Opt. Express **15**, 6744–6749 (2007)
10. Hansen, P.B., Raybon, G., Chien, M.-D., Koren, U., Miller, B.I., Young, M.G., Verdiell, J.-M., Burrus, C.A.: A 1.54-μm monolithic semiconductor ring laser: CW and mode-locked operation. IEEE Photonics Technol. Lett. **4**, 411–413 (1992)
11. Sorel, M., Giuliani, G., Scirè, A., Miglierina, R., Donati, S., Laybourn, P.J.R.: Operating regimes of GaAs–AlGaAs semiconductor ring lasers: experiment and model. IEEE J. Quantum Electron. **39**, 1187–1195 (2003)
12. Sorel, M., Laybourn, P.J.R., Scirè, A., Balle, S., Giuliani, G., Miglierina, R., Donati, S.: Alternate oscillations in semiconductor ring laser. Opt. Lett. **27**, 1992–1994 (2002)
13. Giuliani, G., Miglierina, R., Sorel, M., Scirè, A.: Linewidth, autocorrelation, and cross-correlation measurements of counterpropagating modes in GaAs-AlGaAs semiconductor ring lasers. IEEE J. Sel. Top. Quantum Electron. **11**, 1187–1192 (2005)
14. Sorel, M., Laybourn, P.J.R., Giuliani, G., Donati, S.: Unidirectional bistability in semiconductor waveguide ring lasers. Appl. Phys. Lett. **80**, 3051–3053 (2002)
15. Cao, H., Ling, H., Liu, C., Deng, H., Benavidez, M., Smagley, V.A., Caldwell, R.B., Peake, G.M., Smolyakov, G.A., Eliseev, P.G., Osiński, M.: Large S-section-ring-cavity diode lasers: directional switching, electrical diagnostics, and mode beating spectra. IEEE Photonics Technol. Lett. **17**, 282–284 (2005)
16. Cao, H., Deng, H., Ling, H., Liu, C., Smagley, V.A., Caldwell, R.B., Smolyakov, G.A., Gray, A.L., Lester, L.F., Eliseev, P.G., Osiński, M.: Highly unidirectional InAs/InGaAs/GaAs quantum-dot ring lasers. Appl. Phys. Lett. **86**, 203117-1–203117-3 (2005)
17. Sohler, W., Das, B.K., Dey, D., Reza, S., Suche, H., Ricken, R.: Erbium-doped lithium niobate waveguide lasers. IEICE Trans. Electron. **E88-C**, 990–997 (2005)
18. Das, B.K., Suche, H., Sohler, W.: Single-frequency Ti:Er:LiNbO$_3$ distributed Bragg reflector waveguide laser with thermally fixed photorefractive cavity. Appl. Phys. B **73**, 439–442 (2001)
19. Das, B.K., Ricken, R., Sohler, W.: Integrated optical distributed feedback laser with Ti:Fe:Er:LiNbO$_3$ waveguide. Appl. Phys. Lett. **82**, 1515–1517 (2003)
20. Sohler, W., Das, B., Reza, S., Ricken, R.: Recent progress in integrated rare-earth doped LiNbO$_3$ waveguide lasers. In: Proceedings of the 9th Optoelectronics and Communications Conference, Kanagawa, Japan, 12–16 July 2004, p. 568
21. Vossler, L., Olinger, M.D., Page, J.L.: Solid medium optical ring laser rotation sensor. US Patent # 5,408,492 (1995)
22. Rong, H., Xu, S., Kuo, Y.-H., Sih, V., Cohen, O., Raday, O., Paniccia, M.: Low-threshold continuous-wave Raman silicon laser. Nat. Photonics **1**, 232–237 (2007)
23. De Leonardis, F., Passaro, V.M.N.: Modeling and performance of a guided-wave optical angular-velocity sensor based on Raman effect in SOI. J. Lightwave Technol. **25**, 2352–2366 (2007)
24. Scheuer, J., Green, W.M.J., DeRose, G.A., Yariv, A.: InGaAsP annular Bragg lasers: theory, applications, and modal properties. IEEE J. Sel. Topics Quantum. Electron. **11**, 476–484 (2005)
25. Scheuer, J.: Direct rotation-induced intensity modulation in circular Bragg micro-lasers. Opt. Express **15**, 15053–15059 (2007)

26. Frolov, S.V., Shkunov, M., Fujii, A., Yoshino, K., Vardeny, Z.V.: Lasing and stimulated emission in π-conjugated polymers. IEEE J. Quantum Electron. **36**, 2–11 (2000)
27. Kenji, O.: Semiconductor ring laser gyro. Japanese patent # JP 60,148,185 (1985)
28. Donati, S., Giuliani, G., Sorel, M.: Proposal of a new approach to the electrooptical gyroscope: the GaAlAs integrated ring laser. Alta Frequenza **9**, 61–62 (1997)
29. Armenise, M., Laybourn, P.J.R.: Design, simulation of a ring laser for miniaturised gyroscopes. Proc. SPIE **3464**, 81–90 (1998)
30. Armenise, M.N., Passaro, V.M.N., De Leonardis, F., Armenise, M.: Modeling and design of a novel miniaturized integrated optical sensor for gyroscope applications. J. Lightwave Technol. **19**(10), 1476–1494 (2001)
31. Armenise, M.N., Armenise, M., Passaro, V.M.N., De Leonardis, F.: Integrated optical angular velocity sensor. European Patent # 1219926 (2000)
32. European Space Agency (ESA), IOLG project 1678/02/NL/PA: Ring lasers model and quantum effect in integrated optical angular velocity sensor. Contract Report, December 2003
33. Taguchi, K., Fukushima, K., Ishitani, A., Ikeda, M.: Self-detection characteristics of the Sagnac frequency shift in a mechanically rotated semiconductor ring laser. Measurement **27**, 251–256 (2000)
34. Taguchi, K., Fukushima, K., Ishitani, A., Ikeda, M.: Experimental investigation of a semiconductor ring laser as an optical gyroscope. IEEE Trans. Instrum. Meas. **48**, 1314–1318 (1999)
35. Osiński, M., Cao, H., Liu, C., Eliseev, P.G.: Monolithically integrated twin ring diode lasers for rotation sensing applications. J. Cryst. Growth **288**, 144–147 (2006)
36. Cao, H., Liu, C., Ling, H., Deng, H., Benavidez, M., Smagley, V.A., Caldwell, R.B., Peake, G.M., Smolyakov, G.A., Eliseev, P.G., Osiński, M.: Frequency beating between monolithically integrated semiconductor ring lasers. Appl. Phys. Lett. **86**, 041101-1–041101-3 (2005)
37. Osiński, M., Taylor, E.W., Eliseev, P.G.: Monolithically integrated semiconductor unidirectional ring laser rotation sensor/gyroscope. US Patent # 6,937,342 (2005)
38. Lawrence, A.W.: Thin film laser gyro. US Patent # 4,326,803 (1982)
39. Haavisto, J.R.: Passive resonant optical microfabricated inertial sensor and method using same. US Patent # 5,872,877 (1999)
40. Rabus, D.G.: Integrated Ring Resonators: The Compendium. Springer, Berlin (2007)
41. Ciminelli, C., Dell'Olio, F., Campanella, C.E., Passaro, V.M.N., Armenise, M.N.: Integrated optical ring resonators: modelling and technologies. In: Emersone, P.S. (ed.) Progress in Optical Fibers. Nova Publisher, New York (2009)
42. Ciminelli, C., Campanella, C.E., Armenise, M.N.: Optimized design of integrated optical angular velocity sensors based on a passive ring resonator. J. Lightwave Technol. **27**, 2658–2666 (2009)
43. Ciminelli, C., Passaro, V.M.N., Dell'Olio, F., Armenise, M.N.: Quality factor and finesse optimization in buried InGaAsP/InP ring resonators. J. Eur. Opt. Soc. Rap. Publ. **4**, 09032 (2009)
44. Adar, R., Serbin, M.R., Mizrahi, V.: Less than 1 dB per meter propagation loss of silica waveguides measured using a ring resonator. J. Lightwave Technol. **12**, 1369–1372 (1994)
45. Ciminelli, C., Peluso, F., Armenise, M.N.: A new integrated optical angular velocity sensor. Proc. SPIE **5728**, 93–100 (2005)
46. Ciminelli, C.: Innovative photonic technologies for gyroscope systems. In: EOS Topical Meeting—Photonic Devices in Space, Paris, 18–19 October 2006 (Invited Paper)
47. Ciminelli, C., Peluso, F., Armandillo, E., Armenise, M.N.: Modeling of a new integrated optical angular velocity sensor. In: Optronics Symposium (OPTRO), Paris, 8–12 May 2005
48. Ciminelli, C., Peluso, F., Catalano, N., Bandini, B., Armandillo, E., Armenise, M.N.: Integrated optical gyroscope using a passive ring resonator. In: ESA Workshop, Noordwijk, The Netherlands, 3–5 October 2005

49. European Space Agency (ESA), IOLG project 1678/02/NL/PA: Semiconductor optical amplifier modelling. Contract report, November 2004
50. European Space Agency (ESA), IOLG project 1678/02/NL/PA: Passive resonant angular velocity sensor modelling. Contract report, November 2004
51. European Space Agency (ESA), IOLG project 1678/02/NL/PA. Final report, December 2008
52. Suzuki, K., Takiguchi, K., Hotate, K.: Monolithically integrated resonator microoptic gyro on silica planar lightwave circuit. J. Lightwave Technol. **18**, 66–72 (2000)
53. Ma, H., Zhang, X., Jin, Z., Ding, C.: Waveguide-type optical passive ring resonator gyro using phase modulation spectroscopy technique. Opt. Eng. **45**, 080506 (2006)
54. Ma, H., Wang, S., Jin, Z.: Silica waveguide ring resonators with multi-turn structure. Opt. Commun. **281**, 2509–2512 (2008)
55. Sohler, W., Hu, H., Ricken, R., Quiring, V., Vannahme, C., Herrmann, H., Büchter, D., Reza, S., Grundkötter, W., Orlov, S., Suche, H., Nouroozi, R., Min, Y.: Integrated optical devices in lithium niobate. Opt. Photon. News **19**(1), 24–31 (2008)
56. West, B.: Ion-exchanged glass waveguides. In: Gupta, M., Ballato, J. (eds.) The Handbook of Photonics. CRC Press, Boca Raton (2007)
57. Li, G., Winick, K.A., Youmans, B.R., Vikjaer, E.A.J.: Design, fabrication and characterization of an integrated optic passive resonator for optical gyroscopes. In: Proceedings of the Institute of Navigation's 60th Annual Meeting, Dayton, USA, 7–9 June 2004
58. Duwel, A., Barbour, N.: MEMS development at Draper Laboratory. In: SEM Annual Conference, Charlotte, USA, 2–4 June 2003
59. Hsiao, H., Winick, K.A.: Planar glass waveguide ring resonators with gain. Opt. Express **15**, 17783–17797 (2007)
60. Vannahme, C., Suche, H., Reza, S., Ricken, R., Quiring, V., Sohler, W.: Integrated optical Ti:LiNbO3 ring resonator for rotation rate sensing. In: Proceedings of the European Conference on Integrated Optics (ECIO), Copenhagen, Denmark 2007, paper WE1, 25–27 April 2007
61. Grover, R., Absil, P.P., Ibrahim, T.A., Ho, P.-T.: III-V semiconductor optical micro-ring resonators. In: Michelotti, F., Driessen, A. Bertolotti, M. (eds.) Microresonators Building Bloks for VLSI Photonics. American Institute of Physics (2004)
62. Rabus, D.G., Hamacher, M., Troppenz, U., Heidrich, H.: Optical filters based on ring resonators with integrated semiconductor optical amplifiers in GaInAsP–InP. IEEE J. Sel. Top. Quantum Electron. **8**, 1405–1411 (2002)
63. Choi, S.J., Peng, Z., Yang, Q., Hwang, E.H., Dapkus, P.D.: A high-Q wavelength filter based on buried heterostructure ring resonators integrated with a semiconductor optical amplifier. IEEE Photonics Technol. Lett. **17**, 2101–2103 (2005)
64. Choi, S.J., Djordjev, K., Peng, Z., Yang, Q., Choi, S.J., Dapkus, P.D.: Laterally coupled buried heterostructure high-Q ring resonators. IEEE Photonics Technol. Lett. **16**, 2266–2268 (2004)
65. Ciminelli, C., Passaro, V.M.N., Dell'Olio, F., Armandillo, E., Armenise, M.N.: Three-dimensional investigation of scattering loss in InGaAsP-InP and Silica-on-Silicon bent waveguides. J. Eur. Opt. Soc. Rap. Publ. **4**, 09015 (2009)
66. Ciminelli, C., Dell'Olio, F., Passaro, V.M.N., Armenise, M.N.: Low-loss InP-based ring resonators for integrated optical gyroscopes. In: Caneus 2009 Workshop, NASA Ames Center, Moffett Field, CA, USA, 1–6 March 2009
67. Xia, F., Sekaric, L., Vlasov, Y.: Ultracompact optical buffers on a silicon chip. Nat. Photonics **1**, 65–71 (2007)
68. Yariv, A., Xu, Y., Lee, R.K., Scherer, A.: Coupled-resonator optical waveguide: a proposal and analysis. Opt. Lett. **24**, 711–713 (1999)
69. Heebner, J.E., Boyd, R.W., Park, Q.-H.: SCISSOR solitons and other novel propagation effects in microresonator-modified waveguides. J. Opt. Soc. Am. B **19**, 722–731 (2002)

70. Heebner, J.E., Chak, P., Pereira, S., Sipe, J.E., Boyd, R.W.: Distributed and localized feedback in microresonator sequences for linear and nonlinear optics. J. Opt. Soc. Am. B **21**, 1818–1832 (2004)
71. Peng, C., Li, Z., Xu, A.: Optical gyroscope based on a coupled resonator with the all-optical analogous property of electromagnetically induced transparency. Opt. Express **15**, 3864–3875 (2007)
72. Peng, C., Li, Z., Xu, A.: Rotation sensing based on a slow-light resonating structure with high group dispersion. Appl. Opt. **46**, 4125–4131 (2007)
73. Leonhardt, U., Piwnicki, P.: Ultrahigh sensitivity of slow-light gyroscope. Phys. Rev. A **62**, 55801 (2000)
74. Scheuer, J., Yariv, A.: Sagnac effect in coupled-resonator slow-light waveguide structures. Phys. Rev. Lett. **96**, 53901 (2006)
75. Zhang, Y., Wang, N., Tian, H., Wang, H., Qiu, W., Wang, J., Yuan, P.: A high sensitivity optical gyroscope based on slow light in coupled-resonator-induced transparency. Phys. Lett. A **372**, 5848–5852 (2008)
76. Matsko, A.B., Savchenkov, A.A., Ilchenko, V.S., Maleki, L.: Optical gyroscope with whispering gallery mode optical cavities. Opt. Commun. **233**, 107–112 (2004)
77. Steinberg, B.Z., Scheuer, J., Boag, A.: Rotation-induced superstructure in slow-light waveguides with mode-degeneracy: optical gyroscopes with exponential sensitivity. J. Opt. Soc. Am. B **24**, 1216–1224 (2007)
78. Joannopoulos, J.D., Meade, R.D., Winn, J.N.: Photonic Crystals-Molding the Flow of Light. Princeton University Press (1995)
79. Krauss, F., de la Rue, R.M., Brand, S.: Two-dimensional photonic bandgap structures operating at near-infrared wavelengths. Nature **383**, 699–702 (1996)
80. Steinberg, B.Z., Boag, A.: Splitting of microcavity degenerate modes in rotating photonic crystals-the miniature optical gyroscopes. J. Opt. Soc. Am. B **24**, 142–151 (2007)
81. Shamir, A., Steinberg, B.Z.: On the electrodynamics of rotating crystals, micro-cavities, and slow-light structures: from asymptotic theories to exact green's function based solutions. In: Proceedings of the International Conference on Electromagnetics in Advanced Applications, Torino, Italy, 17–21 Sept 2007, pp. 45–48

Chapter 6
MEMS Gyroscopes

Micro-Electro-Mechanical Systems (MEMS) are very attractive devices which are exploited in a wide spectrum of application fields such as automotive, consumer electronics, medicine, biotechnology and so on. Global market for MEMS is quickly growing and it is expected to exceed 9 billion dollars in 2010 [1].

One of the most innovative applications of MEMS is in the field of space systems and crafts. MEMS technology has enabled the development of RF switches and phase shifters for spacecraft communications, lab-on-a-chip microsensors for remote chemical detection, highly miniaturized science instruments, compact thermal control systems for pico- and nano-satellites, and inertial sensors for spacecraft navigation [2]. MEMS space applications have been firstly envisaged in late 1990s and an intense research effort has been developed to fabricate reliable prototypes, so that a quick development of MEMS space applications can be predicted in the next few years.

Silicon-based MEMS inertial sensors are commercially available devices [3]. In particular, MEMS gyroscopes with their reduced cost, size, weight, and power consumption have a lot of applications especially in automotive and consumer electronics [4]. Automotive industry is an important commercial driver for MEMS gyro development. Currently, cars are equipped with a number of control and safety systems, such as traction control systems, rollover detection units and antilock braking systems employing MEMS angular rate sensors with a resolution around 0.1 °/s (=360 °/h). Gaming consoles, anti-jitter systems for digital cameras, wireless 3D pointing devices are typical examples of electronic equipments using small and low-cost MEMS gyros. Short-range navigation and guidance of autonomous vehicles in the absence of the GPS signal and, in general, GPS-systems augmentation are other potential application fields for MEMS gyros having a minimum detectable angular rate less than 10 °/h.

Since they use vibrating mechanical elements to sense rotation, MEMS gyros are vibratory gyros. Vibratory angular rate sensors with macroscopic dimensions in the range of centimetres are under development since the 1950s and they are typically based on a stable quartz resonators [5]. The best performing quartz

M N Armenise et al., *Advances in Gyroscope Technologies*,
DOI: 10.1007/978-3-642-15494-2_6, © Springer-Verlag Berlin Heidelberg 2010

vibratory gyroscope is the hemispherical resonator gyro (HRG), which exhibits a bias stability less than 0.001 °/h, and a resolution less than 0.03 °/h (assuming the sensor bandwidth = 20 Hz). This expensive sensor is based on a resonator having a diameter around 1 cm and thus is not a micromachined device. MEMS gyros are miniaturized vibratory gyroscopes typically realized by silicon-based micromachining technology. MEMS angular rate sensors based on quartz or metal resonators have also been proposed. A typical example of quartz MEMS gyro is in [6]. High performance commercial quartz MEMS gyro shows a bias drift of 3 °/h, a resolution around 30 °/h, and an ARW = 0.12 °/$\sqrt{h}$. Main drawback of quartz MEMS gyros is that their batch processing is not compatible with fabrication technology of integrated electronic circuits.

MEMS gyroscopes include a mechanical resonator supporting two resonating modes, the primary mode and the secondary mode. As pointed out in Chap. 2, the Coriolis force couples these two resonating modes, i.e. sense and drive modes, when the sensor is subject to angular velocity.

MEMS angular rate sensors can operate in either mode-matched or split-mode condition. Under mode-matched operating conditions, the sense mode is designed to have the same (or nearly the same) resonant frequency as the drive mode. Hence, the rotation-induced Coriolis signal is amplified by the mechanical quality factor of the sense mode. In split-mode condition, the drive and sense modes have two different resonant frequencies. Due to quality factor amplification, gyroscopes operated in mode-matched configuration offer higher sensitivity and better resolution. Split-mode configuration is more common in automotive applications where high robustness is required.

In either split-mode or mode-matched configuration, scale factor of MEMS gyros increases with resonator mass. Then a trade-off between sensor mass and scale factor has to be achieved.

For properly working, the primary mode needs to be excited with a certain amplitude that typically is accurately monitored by a feedback loop. In order to be able to resolve the angular velocity, vibration of the secondary mode has to be detected.

In a MEMS gyroscope, the actuator converts an electrical signal, either voltage or current, into a force that sustains the drive resonant mode. Similarly, the amplitude readout of both resonant modes requires a technique to convert the position of the vibrating mass to a physical quantity which can be measured by an electronic circuit.

Typically primary resonator is actuated by comb electrodes exploiting an electrostatic actuation mechanism. The detection of vibration amplitude is usually performed by capacitive techniques. Alternative actuation mechanisms are piezoelectric or electromagnetic ones. Detection can be also piezoresistive or piezoelectric.

MEMS gyros can be classified according to whether the micromechanical and electronic parts are on a single chip or on two separate chips. The advantages of a single-chip implementation are reduced size and electronic noise due to parasitic capacitances generated at the interconnection between the mechanical and electronic parts. The extra steps required to combine the two dies at the packaging level are also eliminated. On the other hand, two-chip implementation makes possible the separate optimization of the technologies used to realize the mechanical and electronic parts and allows independent yield control of the fabrication processes. The first single-chip MEMS gyro has been commercialized in 2002 [7].

Since noise in MEMS gyros output is due to Brownian motion of the air molecules surrounding the mechanical resonator, high-performance commercially available MEMS angular rate sensors are typically vacuum-packaged to enhance their performance. Also research prototypes are usually characterized in vacuum.

The development of MEMS gyroscopes exhibiting high-performance even at atmospheric pressure would be very welcome to reduce packaging cost and complexity.

Micro-Opto-Electro-Mechanical Systems (MOEMS), which are the result of micro-optics and MEMS technology integration, are innovative micro-systems developed for different applications such as optical switching and sensing [8]. MOEMS gyros are under investigation since several years. Main achievements of this research activity are very briefly reviewed in this chapter.

6.1 Fabrication Technologies

Micromachining technological processes for MEMS/MOEMS gyro fabrication include bulk micromachining, wafer bonding, surface micromachining, electroplating, Lithographie Galvanoformung Abformung (LIGA) and combined surface-bulk micromachining.

In silicon bulk micromachining, microstructures are realized by selectively removing some regions of the substrate. Silicon is removed from the wafer by means of a deep etching process allowing fabrication of three-dimensional structures. Patterned wafers are frequently bonded together, by anodic or fusion bonding, to form more complex structures.

Patterning of Si wafers is performed using various etching techniques. Etching can be wet or dry and isotropic or anisotropic. In wet etching, the wafer is immersed into an etching solution, which removes bulk material from areas not covered by any protective material (SiO_2 or Si_3N_4). Dry etching is performed in a weakly ionized plasma at low pressure. For silicon dry etching, SF_6 plasma is usually used. The depth of etched structures can be controlled by either tuning etching duration or using etch-stop layers. Isotropic etching is usually wet and exhibits the same etching rate in all directions. Of course, anisotropic etching shows different etching rates in different crystallographic directions. It can be either wet or dry. In anisotropic dry etching techniques, such as Deep Reactive Ion Etching (DRIE), the direction perpendicular to the wafer is etched more quickly than the direction parallel to the wafer plane.

Bulk micromachining in quartz has been also utilized to produce MEMS gyros.

Surface micromachining technology allows the fabrication of thin micromechanical devices on the Si wafer surface. Microstructures are formed by growing and etching different kinds of films on top of a silicon wafer. Structural and sacrificial layers are alternatively grown and patterned on the substrate. At the end of the process the sacrificial layers are selectively removed by an etchant, defining the different components of the device.

The most typical structural material is polysilicon, while silicon dioxide is often used as sacrificial material.

Surface micromachining is the preferred technology to fabricate a huge number of MEMS devices simultaneously on a single wafer through only one fabrication run, but realized structures are smaller than those fabricated by bulk micromachining. In order to increase the resonator mass and, consequently, the sensor scale factor thick structural polysilicon layer, grown on a SiO_2 film deposited on the top of the silicon substrate, can be used. An alternative way to fabricate resonators having a large mass is the use of a SOI (Silicon-on-Insulator) wafer. On the top of this wafer there is a single-crystal silicon layer having a typical thickness ranging from a few micrometers to several tens of micrometers. Microstructures are formed in this layer and realized by etching the buried SiO_2 layer grown between the top Si layer and the Si substrate. SOI surface micromachining is emerging as a very attractive fabrication technique for high performance MEMS devices.

Main advantage of surface micromachining is the compatibility with conventional fabrication techniques of integrated circuits. This allows to integrate the MEMS sensing element and the read-out electronic circuit on the same Si chip.

Electroplating is used in MEMS to deposit thick layers of metal. To produce a patterned electroplated layer on silicon, a resist pattern, referred to as a mold, has to be deposited on the substrate. This resist mold is processed by UV-lithography and so structures with a maximum aspect ratio around 10:1 can be realized by this technique.

LIGA makes use of lithography, electrodeposition and molding processes to produce microstructures. A thick layer of X-ray resist with thickness ranging from microns to centimetres is exposed to high-energy X-ray radiation. In this manner, three-dimensional resist structures are obtained. Subsequent electrodeposition fills the resist mold with a metal. After resist removal a free standing metal structure is obtained. The use of a highly collimated X-ray source enables structures with quite vertical sidewalls and aspect ratios larger than 100:1 to be made.

6.2 Research Prototypes and Commercially Available Devices

In the last 20 years, an intense research effort has been paid by a large number of research groups all over the world to propose, fabricate and develop MEMS gyros having a good performance-cost trade-off. MEMS angular rate sensors can be sensitive to rotation around one or two axes. Moreover single-axis gyros can be sensitive to rotation around the axis perpendicular to the substrate on which the device is realized (typically denoted as z-axis) or to angular rate around x- or y-axis which are in the substrate plane. Angular rate sensors sensitive to rotation around x- or y-axis are called lateral axis gyros.

In this paragraph z-axis, lateral-axis and dual-axis MEMS gyros are critically reviewed.

6.2.1   z-Axis MEMS Gyros

First MEMS gyroscope fabricated by bulk silicon micromachining has been reported in [9]. This gyroscope, shown in Fig. 6.1, is a two-gimbals structure supported by torsional flexures. The structure, made by heavily p-doped silicon, is undercut and free to move. The outer gimbal is a rectangular frame connected to the supporting substrate by thin beams allowing its rotation around x-axis. Dimensions of the rectangular outer gimbal are 0.35×0.5 mm^2. The inner gimbal is formed within the central region of the outer gimbal and is connected to it by means of torsional flexures. A gold inertial mass having a height of 25 μm is electroplated onto the inner gimbal. The inner gimbal represents a platform that can rotate around the y-axis. The outer gimbal is electrostatically driven into oscillatory motion out of the wafer plane at constant amplitude by using the driving electrodes. The oscillation amplitude is kept constant by automatic gain control. When subjected to a rotation around the axis perpendicular to the wafer plane (z-axis), Coriolis force induces the oscillation of the inner gimbal around y-axis with a frequency equal to 3 kHz. Electrodes over the inner gimbal detect the amplitude of secondary resonating mode. The voltage arising at these electrodes is

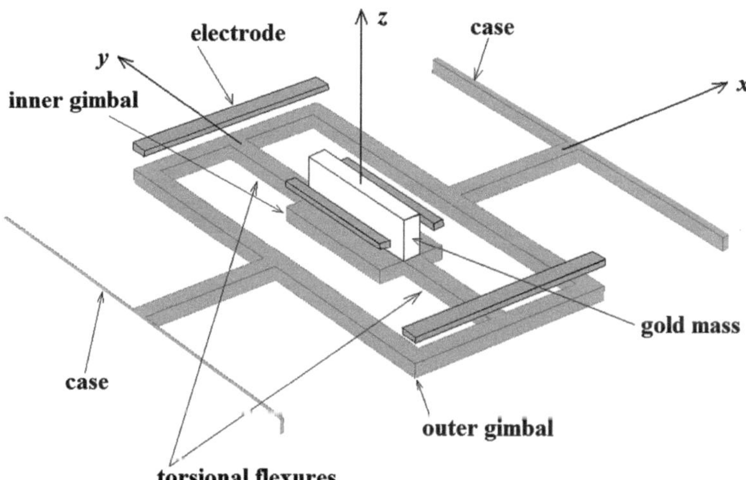

Fig. 6.1   First prototype of MEMS gyro [9]

set to zero by a feedback loop producing an appropriate force that is taken as a measure of the gyro angular rate. A resolution of 4 °/s (=14,400 °/h) with a bandwidth of 1 Hz has been obtained by this gyro having an area of about 0.2 mm^2. A significant improvement has been achieved for this device: bias stability <7 °/h (=2 × 10^{-3} °/s), resolution of 24 °/h or 6.6 × 10^{-3} °/s (assuming a bandwidth of 20 Hz), and ARW = 0.09 °/$\sqrt{h}$ (=1.5 × 10^{-3} °/$\sqrt{s}$) [10].

One of the earliest gyroscopes using a vibrating element is the "gyrotron", which is a tuning fork gyro [11]. This rotation sensor, having dimensions in the range of centimeters, includes two metal tines connected together by a central block. The fork tines oscillates in their plane (primary vibration). When tines are rotated around their axis, a precession around this axis arises. The amplitude of this movement is proportional to angular rate and can be used to sense the tuning fork rotation. The same physical principle has been exploited to realize both z-axis and lateral-axis MEMS gyros.

A z-axis tuning-fork angular rate sensor, realized by a combination of surface and bulk silicon micromachining, has been reported in [12]. This device is based on a two-mass resonator which is electromagnetically stimulated by a permanent magnet mounted on the top of the sensor. Electromagnetic actuation allows an amplitude of primary vibration of 50 μm. Such large amplitude is not achievable by electrostatic actuation and, in principle, enhances the gyro sensitivity. Sense mode amplitude is capacitively detected by comb electrodes.

Very recently, a high-performance tuning fork gyroscope has been developed [13]. The sensor, shown in Fig. 6.2, has been realized using SOI surface micromachining and includes two proof-masses that can oscillate along the two in-plane axes (x- and y-axis). Drive mode is along x-axis and is excited by comb electrodes. Rotation about z-axis induces an energy transfer between the drive

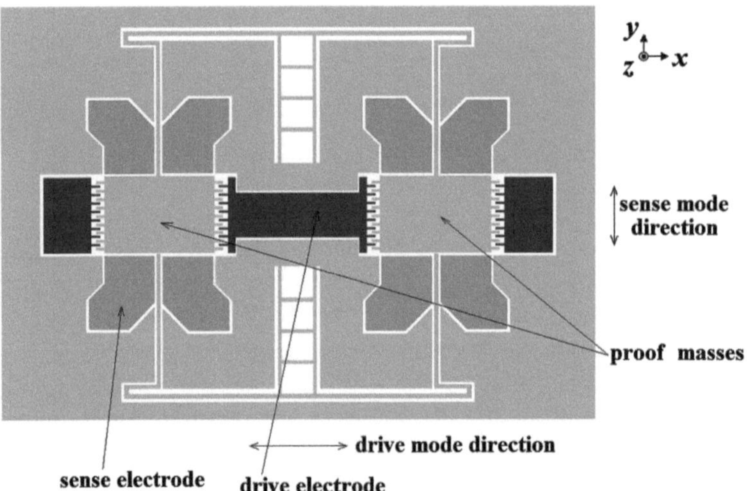

Fig. 6.2 Tuning fork MEMS gyroscope reported in [13]

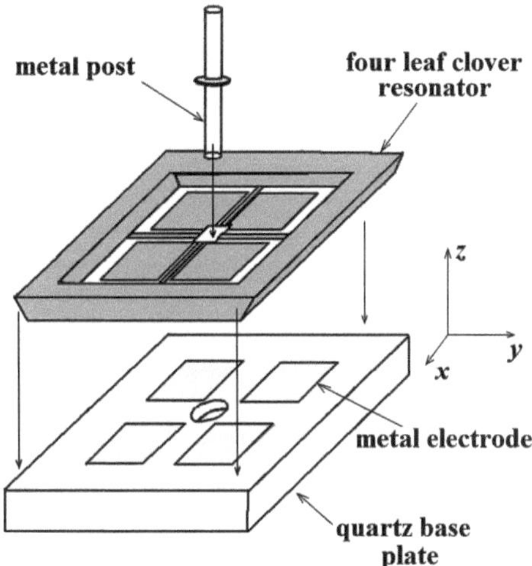

Fig. 6.3 MEMS gyro based on a four leaf clover resonator [15]

mode and the sense one which is directed along y-axis. Sense mode amplitude is detected by four sense electrodes. Sensor area is around 2 mm^2. Details about the CMOS ASIC for gyro read-out are reported in [14]. Characterization of fabricated device was performed in 1-mtorr vacuum and a bias instability of 0.15 °/h (=4 $\times$ 10^{-5} °/s), an ARW = 0.003 °/$\sqrt{h}$ (=5 $\times$ 10^{-5} °/$\sqrt{s}$) and a resolution around 1 °/h or 2.8 $\times$ 10^{-4} °/s (assuming a bandwidth of 20 Hz) were measured.

In 1996 a bulk-micromachined gyroscope significantly enhancing the state-of-the-art was fabricated [15]. The sensor, requiring after-fabrication assembly, resembles a four leaf clover suspended by four thin wires with a metal post attached to the centre (see Fig. 6.3). The silicon clover-leaf structure with the post is bonded to a quartz base plate patterned with gold electrodes. The primary motion is the resonator oscillation about x-axis. When a rotation rate is applied about the z-axis, Coriolis force induces the oscillation of clover-leaf structure about y-axis (secondary motion). The metal post enhances the coupling between drive and sense modes. Electrodes allow to excite the periodic oscillation around x-axis and to sense the amplitudes of both resonating modes. Dimensions of the packaged gyroscope are quite large being in the range of centimetres.

A z-axis MEMS angular rate sensor based on a highly symmetric structure has been reported [16]. The sensor, shown in Fig. 6.4, includes only one proof mass that can oscillate along both in-plane axes (x- and y-axis) and comb electrodes for drive mode excitation and sense mode read-out. The gyroscope is driven into oscillation along x-axis by electrostatic excitation applied between the movable and stationary drive fingers. When an external angular rotation is applied about the z-axis, the oscillation along y-axis is excited. If resonant frequencies of the drive and sense modes are closely matched, the sensor sensitivity is maximized. Drive and sense mode resonance frequencies can be matched by symmetric design of suspension beams. This design strategy usually implies undesired mechanical coupling between drive and sense mode. This problem is prevented by placing the

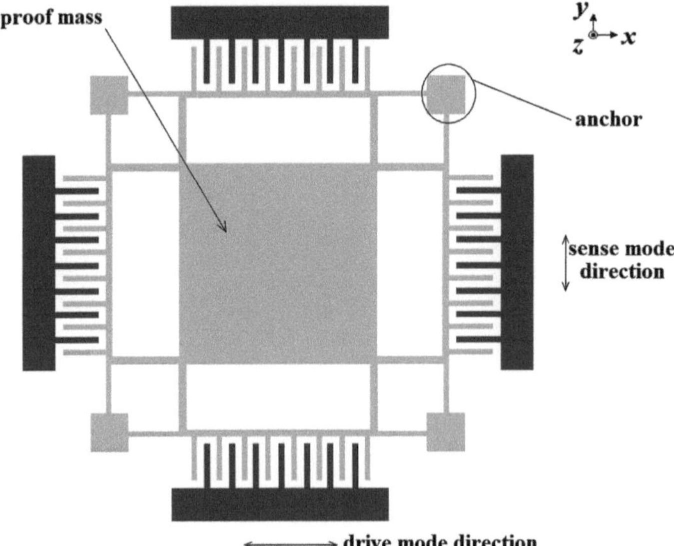

Fig. 6.4 MEMS gyro including only one proof mass [16]

anchors at the outermost corners and connecting them to the movable drive and sense electrodes so that the vibration of drive electrodes does not disturb sense electrodes. The inertial sensor exhibits an ARW around 4×10^{-3} $°/\sqrt{s}$ (=0.2 $°/\sqrt{h}$). More recently, the same research group has proposed a gyro based on the same concept and realized by silicon-on-glass technology exhibiting ARW $= 2 \times 10^{-3}$ $°/\sqrt{s}$ (=0.1 $°/\sqrt{h}$) [17].

In Fig. 6.5 is sketched a polysilicon surface-micromachined MEMS gyroscope [18, 19]. Comb electrodes, located in the middle of the device, have been used to excite the primary vibration which is along x-axis and has a frequency of 12 kHz. Rotation around the axis perpendicular to the substrate (z-axis) excites the sense mode which is along the y-axis. Amplitude of sense mode is capacitively detected by sense electrodes which are on both sides of the device. The amplitude of primary resonance mode is kept constant by automatic gain control. A charge-sensitive amplifier is used for the secondary resonator readout. ARW of this sensor, having an area of 1 mm^2, is equal to 1 $°/\sqrt{s}$ (=60 $°/\sqrt{h}$). In [20] is described a similar z-axis gyro with resonating mass supported by four fishhook-shaped springs. This last sensor exhibits a resolution of 0.1 $°/s$ (=360 $°/h$) with a bandwidth of 2 Hz.

A surface-micromachined MEMS gyro based on resonant sensing was developed about 10 years ago [21]. The sensor consists of a proof mass vibrating in the tens of kilohertz and two resonating sense elements with a designed resonant frequency an order of magnitude higher than that of the proof mass. The architecture of this single-chip gyroscope is shown in Fig. 6.6. Proof mass is suspended by flexures attached to a rigid frame. Comb electrodes excite vibration of the proof

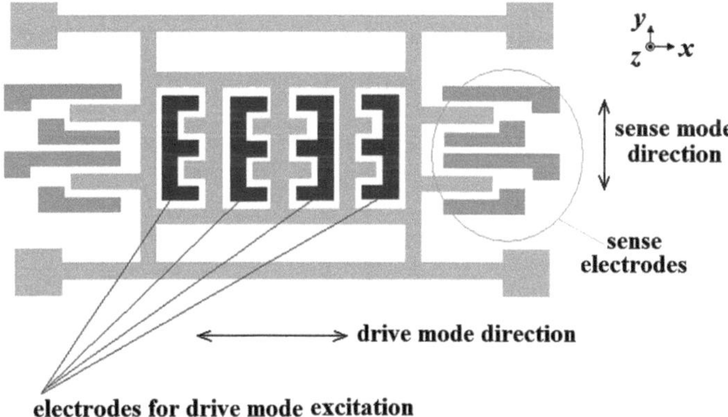

Fig. 6.5 Polysilicon surface-micromachined MEMS gyroscope [18]

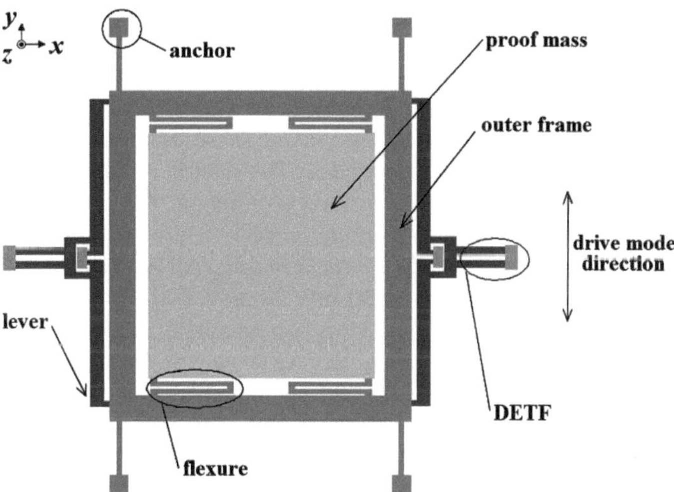

Fig. 6.6 MEMS gyro based on resonant sensing [21]

mass along y-axis (drive mode). If an external rotation around the z-axis is applied
to the chip, the Coriolis force acting on the proof mass is transmitted to the outer
frame. A lever mechanism amplifies this force prior to its being communicated
axially onto two double-ended tuning fork (DETF) resonators placed on both sides
of the outer frame. The periodic compression and tension of tines included in the
DETF resonator due to Coriolis force modulates the resonance frequency of the
two DETF resonators. By monitoring the oscillation frequency of these two resonators, the rotation rate applied to the device can be estimated. ARW = 0.3 $^\circ/\sqrt{s}$
(=18 $^\circ/\sqrt{h}$) was evaluated for this gyro.

A single-chip surface-micromachined MEMS gyro having quite good performance and low cost was also developed in recent years [22]. The sensor includes

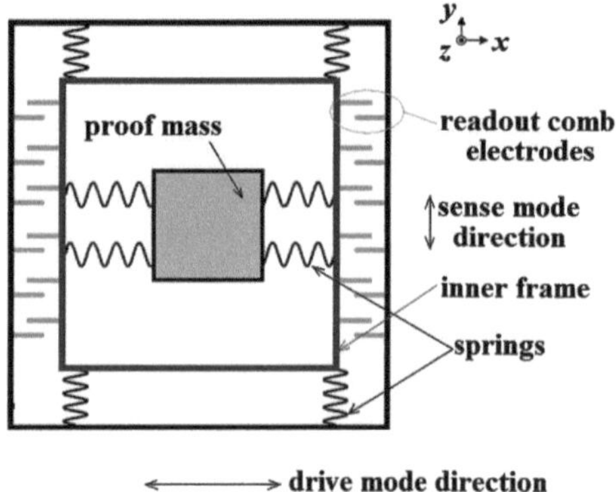

Fig. 6.7 Principle scheme of ADI MEMS gyros [22]

two mechanically independent resonators that differentially sense signals due to angular rate and reject common-mode external disturbances that are unrelated to angular motion. The principle scheme of each resonator is shown in Fig. 6.7. The proof mass vibration along x-axis is induced by comb electrodes. Gyro rotation causes the inner frame vibration along y-axis. Sense mode amplitude is capacitively sensed by comb readout electrodes. The device is fabricated by a 3-μm BiCMOS process with 4-μm thick structural polysilicon. The system includes self-test electrodes to ensure that it is working properly. Gyro resolution, measured at atmospheric pressure, is around 0.015 °/s (=54 °/h), ARW is equal to 0.003 $°/\sqrt{s}$ (=0.2 $°/\sqrt{h}$), power consumption is 30 mW, weight is 0.3 g and operating temperature range is from $-55°C$ to $+85°C$. A commercial digital-output MEMS gyro based on the same operating principle has ARW $= 0.56$ $°/\sqrt{h}$ (=0.01 $°/\sqrt{s}$), bias stability equal to 0.0016 °/s (=6 °/h) and resolution of 0.04 °/s (=144 °/h) [23]. Price of this sensor is larger than $ 500 per unit in thousand-piece quantities.

A high-performance z-axis MEMS gyroscope in which the drive and sense modes are effectively decoupled has been reported in [24]. SOI surface micromachining has been used for sensor fabrication. In this angular rate sensor, shown in Fig. 6.8, the outer frame is suspended by drive mode springs in such a way that it can only move along the x-direction. Comb electrodes are used to induce this vibratory motion. The primary vibration is transferred to an inner mass by sense mode springs. These springs are very stiff along the drive mode direction (x-axis) but they bend easily along the sense direction (y-axis). Because of rotation around z-axis, the inner mass experiences Coriolis force directed along y-axis. This Coriolis force acts upon the mass-spring system formed by the inner mass and the sense springs. By detecting the acceleration along y-axis, the angular velocity can be measured. ARW of this sensor is about 0.04 $°/\sqrt{h}$ (=7 $\times$ 10^{-4} $°/\sqrt{s}$).

Vibrating ring gyroscopes represent a very attractive class of z-axis MEMS angular rate sensors. Good performance of these gyros, demonstrated by different research groups, is mainly due to the symmetry of the resonating element. In these sensors the resonator has a ring structure with a diameter of the order of a few

Fig. 6.8 z-Axis MEMS
gyroscope reported in [24]

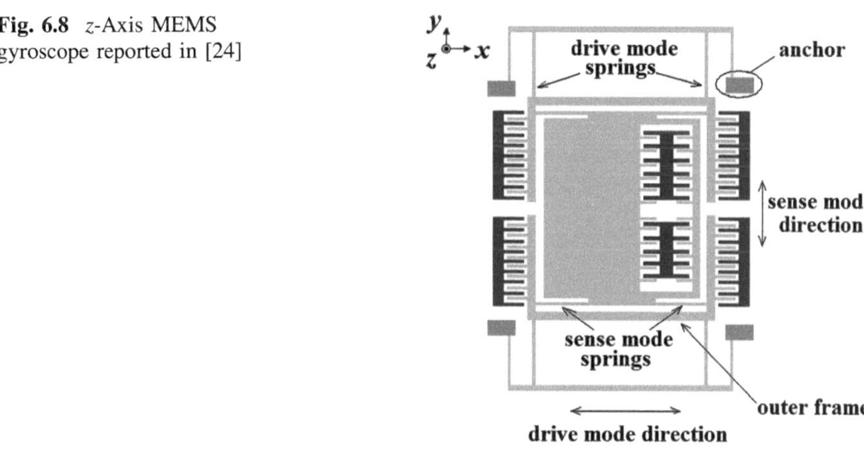

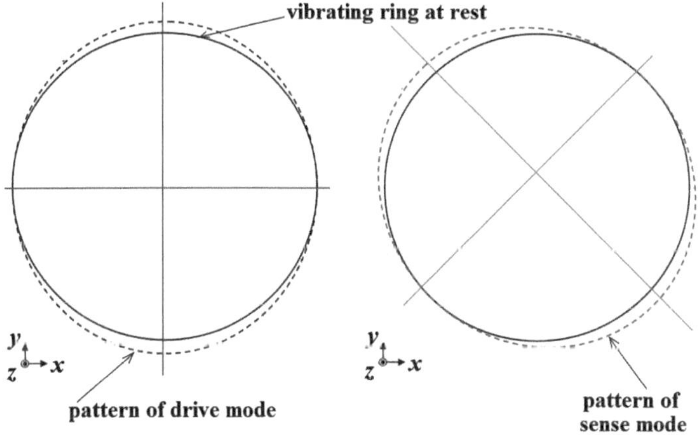

Fig. 6.9 Basic structure of vibrating ring MEMS gyroscopes

millimeters. The primary vibration of the ring, excited by the actuation mechanism, forms an elliptically shaped pattern with the two axes along x and y
directions. Any rotation around the axis normal to the ring structure transfers
energy to a secondary vibration mode (sense mode). Secondary vibration of the
ring forms an elliptically shaped pattern having the two axes rotated of 45° with
respect to x and y axes. The amplitude of secondary vibration is accurately
monitored to estimate the gyro angular rate. The symmetry of the structure provides identical drive and sense resonance frequencies. This enhances gyro sensitivity. Elliptically shaped patterns formed by primary and secondary vibrations are
shown in Fig. 6.9.

The first micromachined vibrating ring gyroscope was reported in 1994 [25]. It
is based on a nickel electroplated ring structure which is fabricated on a wafer

including standard CMOS circuitry for sensor readout and primary mode excitation. Primary resonant mode is electrostatically excited. When the device rotates around z-axis, energy transfer from the primary mode to the secondary one occurs. Amplitude of sense mode is capacitively detected. The sensor exhibits a resolution of 0.5 °/s (=1800 °/h) with a bandwidth of 10 Hz and an ARW = 0.16 °/$\sqrt{s}$ (=9.6 °/$\sqrt{h}$). More recently, an improved vibrating ring gyroscope based on a high aspect ratio polysilicon ring has been reported by the same research group [26]. The ring structure, realized by deep etching a silicon substrate and refilling the trenches with polysilicon deposited over a sacrificial SiO_2 layer, has a diameter of 1.1 mm, a width of 4 μm and a height of 80 μm. Sixteen electrodes are located along the ring to excite the primary vibration and to sense the secondary one. Gap between the ring and the electrodes is 1.4 μm wide. A post having a diameter of 0.12 mm and eight springs are used to support the ring and to anchor it to the silicon substrate. Drive and sense modes resonate at 29 kHz and gyro resolution is 0.04 °/s (=144 °/h) with a bandwidth of 10 Hz. ARW is equal to 0.01 °/$\sqrt{s}$ (=0.6 °/$\sqrt{h}$).

A commercial MEMS gyroscope based on a bulk micromachined silicon ring sensing element having a diameter of a few mm is available [27]. The mechanical resonator, fabricated by DRIE, is connected by eight spokes to a support frame. Current-carrying conductor loops are deposited on the surface of the ring structure. These loops and the magnetic field generated by a permanent magnet excite the drive mode. Then the vibratory gyro uses magnetic actuation and detection and so it cannot be further miniaturized. Best performing version of this sensor exhibits a bias drift of 3 °/h (=0.0008 °/s), a sensitivity around 30 °/h (=0.008 °/s) and an ARW of 0.1 °/$\sqrt{h}$ (=0.0015 °/$\sqrt{s}$).

In 2002 market and feasibility studies founded by ESA about the possibility to develop a low-cost European MEMS gyro suitable for space applications with resolution in the range 1–10 °/h were completed. Reliability testing has been performed on a commercially available MEMS ring gyro [28].

The European Silicon MEMS Rate Sensor (SiREUS) project founded by ESA and started in 2005 was finalized to the realization of a low-cost gyro for space application having an ARW < 0.2 °/$\sqrt{h}$ and a bias stability around 5–10 °/h as objective. Last published results report a bias stability in the range of 10–20 °/h (0.003–0.005 °/s) and an ARW = 0.04 °/$\sqrt{h}$ (=7 × 10^{-4} °/$\sqrt{s}$) for the SiREUS prototype [29].

Very recently has been developed an innovative z-axis MEMS gyroscope based on a star mechanical resonator [30]. The operating principle of this sensor is the same as that one utilized by vibrating ring gyroscopes. By means of the star resonator, the area of drive and sense electrodes and resonator mass can be increased with respect to the classical ring resonator. Fabricated gyroscope, characterized in 1-mtorr vacuum, exhibited ARW = 0.09 °/$\sqrt{h}$ (=0.0015 °/$\sqrt{s}$), bias stability of 3.5 °/h (=10^{-3} °/s) and resolution of 25 °/h or 0.007 °/s, assuming a bandwidth of 20 Hz.

6.2.2 *Lateral-Axis MEMS Gyros*

The first tuning-fork lateral-axis MEMS gyroscope has been fabricated on silicon-on-glass substrate [31]. This device, shown in Fig. 6.10, includes two proof masses coupled to each other by a mechanical suspension. The primary motion is the anti-phase vibration of proof masses along x-axis. When the sensor rotates around y-axis, masses vibrate also in the direction perpendicular to the substrate (z-axis) because of the Coriolis force. This is the secondary motion that allows angular rate estimation. The actuation is electrostatic and detection is capacitive. Both actuation and detection are provided by interdigitated comb electrodes. Matching between resonance frequency of drive and sense modes is critical and imposes very stringent fabrication tolerances. Another issue of this sensor is the mechanical quadrature error due to anisoelesticity and other asymmetries in resonator suspension system. Quadrature signal is in phase with the drive signal and phase-shifted of $\pi/2$ with respect to the Coriolis force. Quadrature error is responsible of gyro bias and so, in this device, it is electrostatically compensated. Gyro minimum detectable angular rate is around 1 °/s (=3600 °/h) with a bandwidth of 60 Hz and ARW $= 0.2$ °/$\sqrt{s}$ (=12 °/$\sqrt{h}$).

Another lateral-axis tuning-fork silicon gyro quite similar to that one mentioned above was reported in [32]. In this sensor, sense mode amplitude is not capacitively detected but measured by four piezoresistors.

A surface-micromachined polysilicon gyroscope that is sensitive to angular rate around x- or y-axis was proposed in 1995 [33]. In this device four beams on a silicon substrate support the polysilicon resonator actuated by comb electrodes. When AC electric signal is applied to these electrodes the mass oscillates along x-direction (drive mode). A rotation around y-axis generates a deflection in z direction due to the Coriolis force that is detected as change of the capacitance between the mass and the sense electrode. This device exhibits ARW $= 2$ °/$\sqrt{s}$ (=120 °/$\sqrt{h}$). The same research group has reported a lateral-axis MEMS gyroscope allowing an efficient decoupling of the sense and drive modes [34]. The sensor, fabricated by using SOI surface micromachining, achieves a

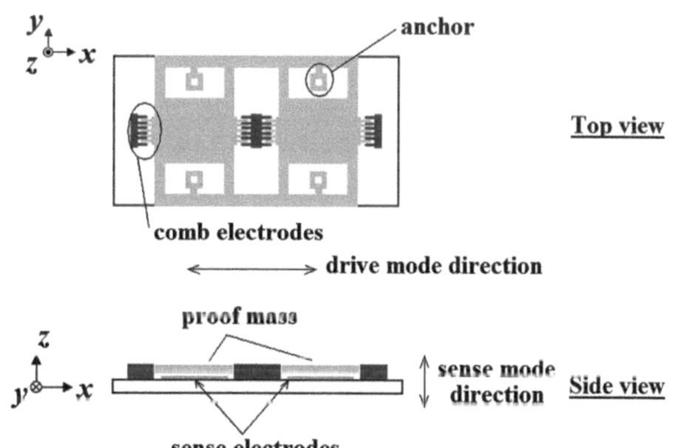

Fig. 6.10 First tuning-fork lateral-axis MEMS gyro [31]

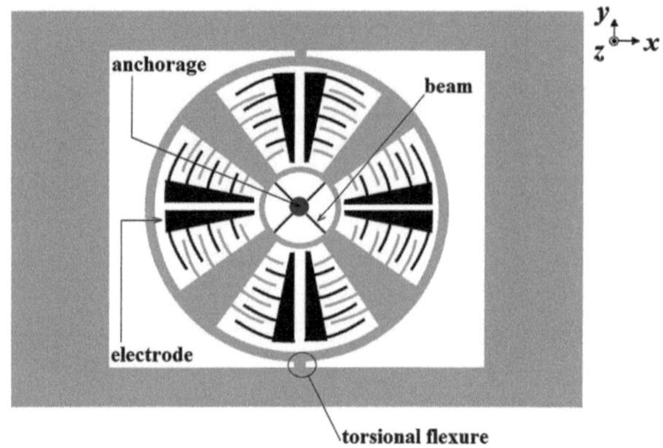

Fig. 6.11 Lateral axis MEMS gyro based on a ring resonating element [37]

resolution of 0.07 °/s (=250 °/h) with bandwidth of 10 Hz and an ARW = 0.02 °/$\sqrt{s}$ (=1.2 °/$\sqrt{h}$).

Other lateral axis gyros exploiting only one proof mass were developed about 5 years ago, with ARW of 0.01 °/$\sqrt{s}$ (=0.6 °/$\sqrt{h}$) [35, 36].

A lateral axis surface-micromachined gyroscope based on a ring resonating element has been reported in [37]. The sensing element, shown in Fig. 6.11, has two decoupled rotary oscillation modes. The drive mode consists of the entire structure rotation around the z-axis. It is electrostatically excited by using the inner spoke electrodes of the inner wheel. The rectangular structure, attached to the inner wheel by torsional springs, exhibits a secondary rotary oscillation about the y-axis (sense mode) in response to rotation about the sensitive axis (x-axis). Rotation about y-axis of the inner wheel is inhibited by the high stiffness of the inner beam suspension. In this device, drive and sense modes are mechanically decoupled and so quadrature error is significantly reduced. The oscillation of the secondary mode is detected capacitively by electrodes on the substrate. The gyro exhibits a resolution of 18 °/h (=0.005 °/s).

6.2.3 Dual-Axis MEMS Gyros

Dual-axis MEMS gyros are capable of sensing angular motion about two axes simultaneously. First dual-axis MEMS angular rate sensor was reported in 1997 [38]. The sensor, shown in Fig. 6.12a, is based on angular resonance of a rigid polysilicon rotor suspended by four torsional springs anchored to the substrate. The inertial rotor is induced to rotate about the z-axis perpendicular to the substrate by comb electrodes. A rotation rate around the x-axis induces a Coriolis angular oscillation around the y-axis and likewise a rotation rate around the y-axis induces a Coriolis angular oscillation around the x-axis. This Coriolis oscillation is measured using the change in capacitance between the rotor and four "quarter circle" electrodes beneath the inertial rotor (see Fig. 6.12b). Dual axes operation can be

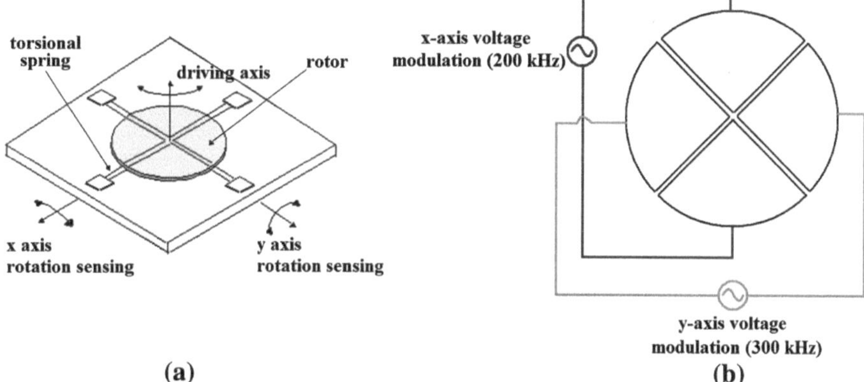

Fig. 6.12 Dual-axis MEMS gyroscope [38]

achieved by using a different modulation frequency for each couple of electrodes. Separate demodulation circuitry for each sense axis provides two voltage outputs proportional to angular rate inputs around respective axes. An ARW of 0.3 $°/\sqrt{s}$ (=18 $°/\sqrt{h}$) with a cross-axis sensitivity ranging from 3 to 16% has been measured.

Also a surface-micromachined dual-axis gyro having a resolution of 0.1 °/s (=360 °/h) with a bandwidth of 5 Hz was fabricated [39].

Dual-axis levitated micromachined spinning-disk gyroscopes have been recently proposed [40]. The sensor is based on a rotor disk, levitated using electromagnetic induction or electrostatic force, that is spun at a very high rate. An attractive electrostatic field helds the disk in a fixed and stable position above the substrate, even if the sensor is tilted or inverted. When the sensor is rotated the disk precesses by an amount proportional to the sensor angular rate.

6.2.4 Summary of MEMS Gyros Performance

Performance of gyros prototypes previously reviewed is summarized in Table 6.1. To compare different gyros, bandwidth has been assumed equal to 20 Hz for all devices. Performance of best performing commercially available MEMS gyros is reported in Table 6.2. From data in the two tables we can conclude that performance of MEMS gyros are not yet compliant with requirements of some very attractive applications such as the inertial navigation, the GPS systems augmentation and the attitude and orbit control of satellites.

6.2.5 Open Issues and Design Criteria

Vibratory gyroscopes can work either in open or closed loop mode to measure the angular velocity. If there is a change in the rotation rate, the amplitude of the sense

Table 6.1 Performance of MEMS gyro prototypes

Authors	Bias instability (°/h)	ARW (°/$\sqrt{h}$)	Resolution in °/h assuming bandwidth = 20 Hz	Fabrication technology
Greiff et al. [9]	–	240	6.4×10^4	Bulk micromachining
Bernstein et al. [31]	–	12	3.2×10^3	Bulk micromachining (silicon-on-glass)
Putty et al. [25]	–	10	2.7×10^3	Nichel electroforming
Tanaka et al. [33]	–	120	3.2×10^4	Poly-Si surface micromachining
Tang et al. [15]	29	1.5	400	Bulk micromachining
Clark et al. [18]	–	60	1.6×10^4	Poly-Si surface micromachining
Mochida et al. [34]	–	1.2	320	SOI, surface micromachining
Ayazi et al. [26]	–	0.6	160	Poly-Si surface micromachining
Geen et al. [22]	–	0.2	50	Poly-Si surface micromachining
Geiger et al. [24]	–	0.04	2.4	SOI, surface micromachining
Xie et al. [35]	31	1.2	320	Bulk micromachining
Kim et al. [36]	–	0.6	160	SOI, surface micromachining
Alper et al. [17]	14.3	0.12	32	Bulk micromachining (silicon-on-glass)
Zaman et al. [13]	0.15	0.003	0.8	SOI, surface micromachining
Zaman et al. [30]	3.5	0.09	24	SOI, surface micromachining

Table 6.2 Performance of best performing commercially available MEMS gyros

Part number	Bias instability (°/h)	ARW (°/$\sqrt{h}$)	Resolution in °/h assuming bandwidth = 20 Hz	Material
HG1900 (Honeywell)	7	0.09	24	Silicon
ADIS16130 (Analog Devices)	6	0.56	150	Silicon
CRS09 (Silicon Sensing)	3	0.1	27	Silicon
QRS116 (Systron Donner)	3	0.12	32	Quartz

mode does not change instantaneously but it required some time to reach the steady state. With matched sense and drive resonant modes, the response time limits the bandwidth of the sensor to a few Hz. The bandwidth of gyroscopes operating in an open-loop mode can be increased with a slight mismatch in the sense and drive mode resonant frequencies but, at the same time, this reduces the sensitivity. In the closed-loop-operating mode, the amplitude of sense mode is continuously monitored and set to zero. This means that the bandwidth and the dynamic range of the sensor can be greater than the corresponding open-loop values even with matched resonant modes. In closed-loop configuration, the bandwidth is limited by the readout and control electronics and can be increased to values approaching the resonant frequency of the structure.

For a z-axis gyroscope driven in the x direction, the amplitude of the vibration due to the Coriolis acceleration (a_y) can be expressed as (see Eq. 6.1):

$$a_y = -\frac{2a_x\omega_x}{\sqrt{\left(\omega_x^2 - \omega_y^2\right)^2 + \omega_x^2\omega_y^2/Q_y^2}}\Omega = -\frac{2a_x\sqrt{mk_x}}{\sqrt{\left(k_x - k_y\right)^2 + \frac{k_xD_y^2}{m}}}\Omega \qquad (6.1)$$

The gyroscope resolution depends on different parameters. In open loop mode with matched sense and drive resonant modes, the resolution can be improved by reducing the noise of the readout circuit, increasing the Coriolis-induced capacitance change of the device, lowering the resonant frequency, increasing the mechanical quality factor and minimizing the parasitic capacitances. Even if a lower resonant frequency of the structure can improve the sensitivity, it must be greater than the environmental noise (>2 kHz). Stronger Coriolis forces can be obtained by increasing the amplitude of drive mode vibration. If the proof mass oscillates in vacuum, very high quality factor can be obtained. Quality factor can be strongly increased by significantly reducing energy losses and it can be achieved when the resonant structure operates in vacuum. This requires hermetically sealed, robust vacuum-packaging techniques, such as those using silicon or glass wafers bonded to the sensor substrate.

Furthermore, if the resonant frequencies of drive and sense modes are matched, the coupling between sense and drive mode is amplified and the resolution is then increased. It is quite difficult to design the device in such a way the two resonance frequencies are perfectly matched over the temperature range and other environmental factors.

A very important performance parameter for a vibratory gyroscope is the bias instability. Imperfections in the geometry of the vibrating mechanical structure, in the electrodes that control the sense and drive modes and asymmetric damping of the structure can produce an output signal even in the absence of rotation. This error, called the quadrature error, can be sometimes orders of magnitude larger than Coriolis signal, and, of course, it may cause errors in rotation rate sensing. Bias instability can be significantly reduced by electrically and mechanically decoupling the sense and drive modes and by reducing the fabrication process errors. Moreover, high-quality materials with low internal damping can also reduce the bias instability.

A high-performance gyroscope should have a stable scale factor, with small temperature sensitivity on a wide dynamic range. To obtain good performance, materials for sensor fabrication have to be chosen with great attention. The use of several materials in the same structure can cause changes of scale factor with the temperature. Then best performance can be obtained in all-silicon devices.

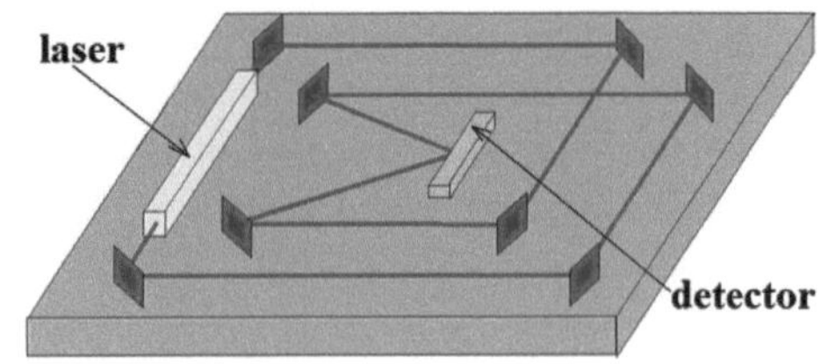

Fig. 6.13 Scheme of the MOEMS gyro reported in [41]

6.3 MOEMS Gyros

MOEMS sensors are under development since several years with the aim of increasing the accuracy of micro-sensors for inertial navigation. It is quite difficult to design a high-performance MOEMS gyroscope because its small dimensions inevitably limits the sensor scale factor.

An interferometric MOEMS gyroscope was proposed in 2000 [41] (see Fig. 6.13). In this device, the basic concept of the interferometric fiber optic gyroscope is integrated with MEMS technology. Micromachined mirrors are placed on a silicon substrate to create a spiral path for the light from the outside, where the laser is placed, to the center of the die, where the interference pattern is detected. The mirrors are arranged in such a way to increase the path length of the device with respect to a standard Sagnac interferometer. The propagation is in free-space and the laser beam only interacts with mirrors on the corners of the spiral path to keep low loss.

Resolution of this MOEMS gyro is limited by the shot noise at the detector. The minimum detectable angular rate is about 0.7 °/s (=2500 °/h).

One year later a hybrid MOEMS gyroscope was proposed [42]. It consists of a MEMS resonator capable to sense the inertial Coriolis force and an optical readout system based on a laser diode that acts as an injection interferometer. The laser beam impinges the resonating structure and the retroreflection is injected back into the laser source. This generates amplitude modulation of the laser power. Authors claims a potential resolution of a few °/h for their gyro.

References

1. Gad-el-Hak, M. (ed.): The MEMS Handbook. CRC Press, Boca Raton (2002)
2. Osiander, R., Darrin M.A.G., Champion, J.L. (eds.): MEMS and Microstructures in Aerospace Applications, CRC Press, Boca Raton (2006)
3. Beeby, S., Ensell, G., Kraft, M., White, N.: MEMS Mechanical Sensors. Artech House, Norwood (2004)
4. Acar, C., Shkel, A.S.: MEMS Vibratory Gyroscopes: Structural Approaches to Improve Robustness. Springer, New York (2008)
5. Lawrence, A.: Modern Inertial Technology: Navigation, Guidance, and Control, Chap. 10. Springer, Berlin (1998)
6. Datasheet of QRS11 Micromachined Angular Rate Sensor by Systron Donner Inertial

7. Geen, J.A.: Progress in integrated gyroscopes. IEEE A&E Systems Magazine, November 2004

8. Motamedi, M.E. (ed.): MOEMS—Micro-Opto-Electro-Mechanical-Systems. SPIE Press, Bellingham (2005)

9. Greiff, P., Boxenhorn, B., King, T., Niles, L.: Silicon monolithic micromechanical gyroscope. In: Proceedings of IEEE International Conference on Solid State Sensors and Actuators, pp. 966–968 (1991)

10. Hanse, J.G.: Honeywell MEMS Inertial Technology & Product Status. In: Proceedings of IEEE PLANS 2004, pp. 43–48 (2004)

11. Lyman, J.: Angular Velocity Responsive Apparatus. US Patent # 2,513,340 (1950)

12. Lutz, M., Golderer, W., Gerstenmeier, J., Marek, J., Maihofer, B., Mahler, S., Münzel, H., Bischof, U.: A precision yaw rate sensor in silicon micromachining. In: Proceedings of International Conference on Solid State Sensors and Actuators, Chicago, USA, pp. 847–850, 16–19 June 1997

13. Zaman, M.F., Sharma, A., Hao, Z., Ayazi, F.: A mode-matched silicon-yaw tuning-fork gyroscope with subdegree-per-hour allan deviation bias instability. IEEE J. Microelectromech. Syst. **17**, 1526–1536 (2008)

14. Sharma, A., Zaman, M.F., Ayazi, F.: A Sub-0.2°/hr bias drift micromechanical silicon gyroscope with sutomatic CMOS mode-matching. IEEE J. Solid State Circuits **44**, 1593–1608 (2009)

15. Tang, T.K., Gutierrez, R.C., Stell, C.B., Vorperian, V., Arakaki, G.A., Rice, J.T., Li, W.J., Chakraborty, I., Shcheglov, K., Wilcox, J.Z., Kaiser, W.J.: A packaged silicon MEMS vibratory gyroscope for microspacecraft. In: Proceedings of the Tenth IEEE International Conference on Micro Electro Mechanical Systems (MEMS '97), Nagoya, Japan, pp. 500–505, 26–30 January 1997

16. Alper, S.E., Akin, T.: A single-crystal silicon symmetrical and decoupled MEMS gyroscope on an insulating substrate. IEEE J. Microelectromech. Syst. **14**, 707–714 (2005)

17. Alper, S.E., Temiz, Y., Akin, T.: A compact angular rate sensor system using a fully decoupled silicon-on-glass MEMS gyroscope. IEEE J. Microelectromech. Syst. **17**, 1418–1429 (2008)

18. Clark, W.A., Howe, R.T., Horowitz, R.: Surface micromachined Z-axis vibratory rate gyroscope. In: Technical Digest. Solid-State Sensor and Actuator Workshop, Hilton Head Island, USA, pp. 283–287, 3–6 June 1996

19. Clark, W.A.: Micromachined z-axis vibratory rate gyroscope. US Patent # 5,992,233 (1999)

20. Park, K.Y., Lee, C.W., Oh, Y.S., Cho, Y.H.: Laterally oscillated and force-balanced micro vibratory rate gyroscope supported by fish hook shape springs. In: Proceedings of IEEE Micro Electro Mechanical Systems Workshop (MEMS '97), Nagoya, Japan, pp. 494–499, 26–30 January 1997

21. Seshia, A.A.: Integrated micromechanical resonant sensors for inertial measurement systems. PhD dissertation, University of California, Berkley, California, USA (2002)

22. Geen, J.A., Sherman, S.J., Chang, J.F., Lewis, S.R.: Single-chip surface micromachined integrated gyroscope with 50°/h Allan deviation. IEEE J. Solid State Circuits **37**, 1860–1866 (2002)

23. Datasheet of ADIS16130 Digital Output. High Precision Angular Rate Sensor by Analog Devices

24. Geiger, W., Butt, W.U., Gaiber, A., Frech, J., Braxmaier, M., Link, T., Kohne, A., Nommensen, P., Sandmaier, H., Lang, W., Sandmaier, H.: Decoupled microgyros and design principle DAVED. Sensors Actuators A **95**, 239–249 (2002)

25. Putty, M.W., Najafi, K.: A micromachined vibrating ring gyroscope. In: Proceedings of the Digest, Solid-State Sensors and Actuators Workshop, Hilton Head Island, USA, pp. 213–220, 13–16 June 1994

26. Ayazi, F., Najafi, K.: A HARPSS polysilicon vibrating ring gyroscope. IEEE J. Microelectromech. Syst. **10**, 169–179 (2001)

27. Datasheet of CRS09 Angular Rate Sensor by Silicon Sensing

28. Barthe, S., Pressecq, F., Marchand, L.: MEMS for space applications: a reliability study. In: 4th Round Table on Micro/Nano Technologies for Space, Noordwijk, The Netherlands, 20–22 May 2003
29. Durrant, D., Dussy, S., Shackleton, B., Malvern, A.: MEMS rate sensors in space becomes a reality. In: AIAA Guidance, Navigation and Control Conference and Exhibit, Honolulu, Hawaii, USA, 18–22 August 2007
30. Zaman, M.F., Sharma, A., Ayazi, F.: The resonating star gyroscope: a novel multiple-shell silicon gyroscope with sub-5 deg/hr Allan deviation bias instability. IEEE Sensors J. **9**, 616–624 (2009)
31. Bernstein, J., Cho, S., King, A.T., Kourepenis, A., Maciel, P., Weinberg, M.: A micromachined comb-drive tuning fork rate gyroscope. In: Proceedings of the IEEE Micro Electro Mechanical Systems Workshop (MEMS '93), Fort. Lauderdale, FL, USA, pp. 143–148, 7–10 February 1993
32. Paoletti, F., Grétillat, M.-A., de Rooij, N.F.: A silicon micromachined vibratory gyroscope with piezoresistive detection and electromagnetic excitation. In: Proceedings of IEEE Micro Electro Mechanical Systems Workshop (MEMS '96), San Diego, USA, pp. 162–167, 11–15 February 1996
33. Tanaka, K., Mochida, Y., Sugimoto, M., Moriya, K., Hasegawa, T., Atsuchi, K., Ohwada, K.: A micromachined vibrating gyroscope. Sensors Actuators A **50**, 111–115 (1995)
34. Mochida, Y., Tamura, M., Ohwada, K.: A micromachined vibrating rate gyroscope with independent beams for the drive and detection modes. In: Proceedings of the IEEE International Conference on Micro-Electro-Mechanical-Systems (MEMS'99), Orlando, USA, pp. 618–623, 16–21 January 1999
35. Xie, H., Fedder, G.K.: Fabrication, characterization, and analysis of a DRIE CMOS-MEMS gyroscope. IEEE Sensors J. **3**, 622–631 (2003)
36. Kim, J., Park, S., Kwak, D., Ko, H., Cho, D.D.: An X-axis single-crystalline silicon microgyroscope fabricated by the extended SBM process. IEEE J. Microelectromech. Syst. **14**, 444–455 (2005)
37. Geiger, W., Merz, J., Fischer, T., Folkmer, B., Sandmaier, H., Lang, W.: The silicon angular rate sensor system DAVED. Sensors Actuators A **84**, 280–284 (2000)
38. Juneau, T., Pisano, A.P., Smith, J.H.: Dual axis operation of a micromachined rate gyroscope. In: Proceedings of the IEEE 1997 International Conference on Solid State Sensors and Actuators (Tranducers '97), Chicago, USA, pp. 883–886, 16–19 June 1997
39. An, S., Oh, Y.S., Lee, B.L., Park, K.Y., Kang, S.J., Choi, S.O., Go, Y.I., Song, C.M.: Dual-axis micro-gyroscope with closed-loop detection. In: Proceedings of the 11th IEEE Micro Electro Mechanical Systems Workshop (MEMS '98), Heidelberg, Germany, pp. 328–333, 25–29 January 1998
40. Damrongsak, B., Kraft, M.: A micromachined electrostatically suspended gyroscope with digital force feedback. In: Proceedings of the IEEE Sensors, Irvine, pp. 401–404, 31 October–3 November 2005
41. Stringer, J.: The air AFIT MEMS interferometric gyroscope (MiG). PhD dissertation, Air Force Institute of Technology, Wright-Patterson Air Force Base, OH, USA (2000)
42. Norgia, M., Donati, S.: Hybrid opto-mechanical gyroscope with injection-interferometer readout. Electron Lett **37**, 756–758 (2001)

Chapter 7
Emerging Gyroscope Technologies

7.1 Performance of Commercial Gyroscopes

Nowadays angular rate sensors commercially available are He–Ne ring laser gyros, interferometric fiber optic gyros, hemispherical resonant gyros, spinning mass gyros (SMGs), and MEMS gyros. High performance gyro market is surely dominated by the He–Ne RLG which is widely used in aeronautic industry. MEMS gyros are exploited in low cost applications, which do not require too high performance. Fiber optic gyros, covering a very wide performance range, are used for space applications such as attitude and orbit control and rover vehicle navigation by NASA, ESA and Japan Aerospace Exploration Agency. European Space Agency has supported the development of a medium performance HRG to be used in future space missions. MEMS gyros for space industry are under development since several years but achieved performance is not fully compliant with the requirements of most of space applications. Space qualification of MEMS gyros has not yet been completed.

Performance of commercially available gyros is summarized in Fig. 7.1.

Integrated microphotonics is a well established technological platform. In the last few decades it has demonstrated several advantages with respect to competing technologies and it has allowed the development of new generation access and transport telecom networks. A very significant impact of the integrated optics also on gyro technology in the next few years is expected. Miniaturization, accuracy, high reliability and immunity to external disturbances allowed by integrated photonics can lead to the medium-term development of a new generation of angular rate sensors. InP-based photonic integrated circuits technology may enable the fabrication of fully integrated gyroscopes which could compete with quality MEMS angular rate sensors in the medium performance gyro market.

Integrated optical gyros are expected to have shortly a resolution in the range 1–10 °/h and a bias drift less than 0.5 °/h.

M. N. Armenise et al., *Advances in Gyroscope Technologies*,
DOI: 10.1007/978-3-642-15494-2_7, © Springer-Verlag Berlin Heidelberg 2010

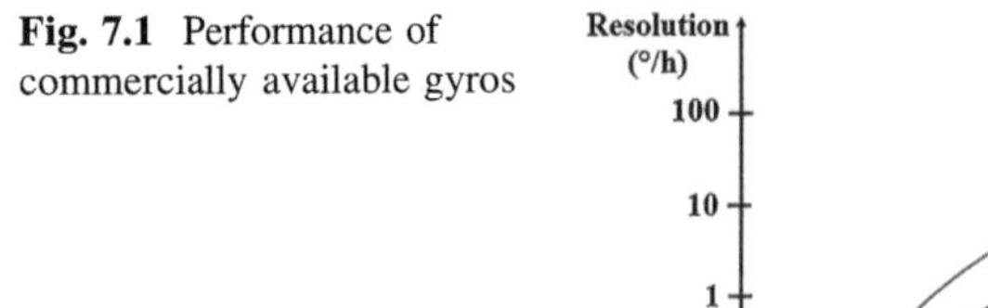

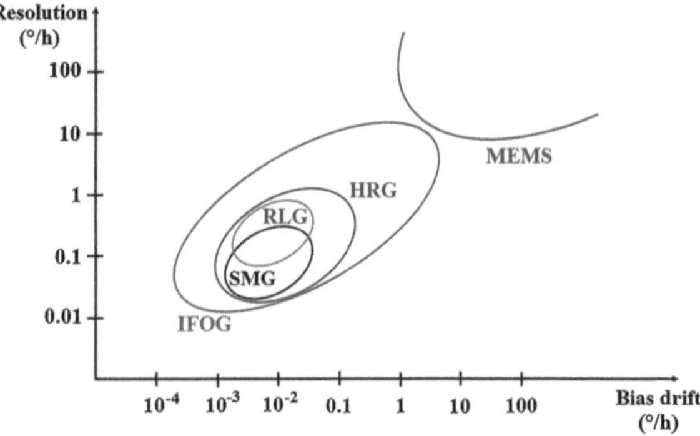

Fig. 7.1 Performance of commercially available gyros

7.2 New Concepts of Gyros

In the last few decades innovative technologies have been both theoretically and experimentally explored for gyroscopic applications. These technologies are usually very complex and expensive and so they are applied in very specific scientific fields such as geodesy and general relativity. In this paragraph the most attractive emerging technologies for angular rate measurements are introduced and briefly reviewed.

7.2.1 Nuclear Magnetic Resonance Gyroscope

Nuclei of some isotopes exhibit a non-zero overall spin angular moment and a magnetic moment parallel to it. If these nuclei are exposed to an external magnetic field not parallel to the magnetic moment, a precession of the spin axis about the direction of the external magnetic field $\mathbf{B_0}$ is observed [1]. This precession, known as Larmor precession, has a characteristic frequency given by:

$$\omega_{NMR} = \gamma_g \mathbf{B_0} \tag{7.1}$$

where γ_g is the gyromagnetic ratio, i.e. the ratio between the angular moment and the magnetic moment and depends on the specific isotope. The frequency ω_{NMR} is called nuclear magnetic resonance (NMR) frequency and can be measured by different techniques, including optical ones.

Magnitude of the magnetic moment of a single nucleus is very small and, at thermal equilibrium, a random orientation of moments is established in an ensemble of atoms. Different techniques can be used to orient a significant fraction of nuclear magnetic moments along a specific direction in a set of atoms. In this manner Larmor precession can be observed and NMR frequency can be measured.

NMR frequency is sensitive to rotation and then NMR gyroscopes are based on the measure of the NMR frequency shift due to rotation [2]. Since NMR frequency

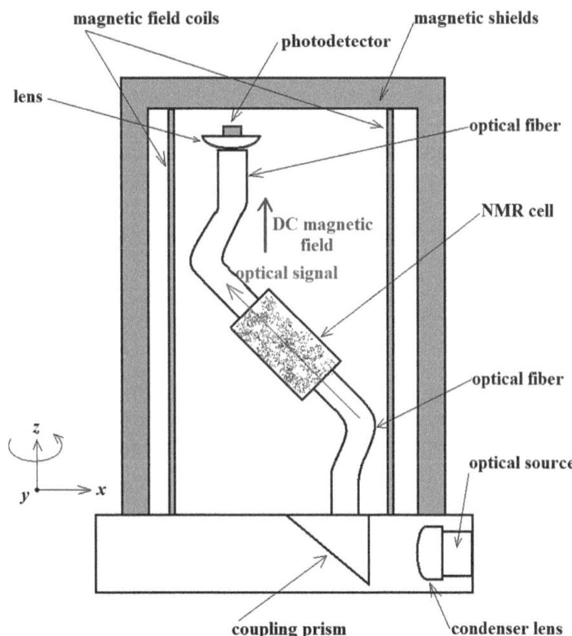

Fig. 7.2 Architecture an NMR gyro [3]

depends on the applied magnetic field, usually at least two spin species with different γ_g values are used in an NMR gyro to cancel this dependence.

At the end of the 1970s two NMR gyros based on two different couples of isotopes (^{129}Xe/^{83}Kr and ^{199}Hg/^{201}Hg) have been patented [3, 4]. Basic structure of the angular rate sensor described in [3] is shown in Fig. 7.2. The sensor is sensitive to rotation around z-axis and its main elements are the NMR cell, containing rubidium vapors, ^{129}Xe and ^{83}Kr, the optical source, the photodetector, a set of magnetic shields and a set of magnetic field coils. Magnetic shields avoid that external magnetic fields perturb the magnetic field generated by magnetic field coils. These coils generate a DC magnetic field directed along the z-axis and two AC magnetic fields directed along x and y axes.

The NMR cell is illuminated by a circularly polarized optical signal. The component of this signal along the z-axis induces the alignment of Rb atoms magnetic moments along the direction parallel to that of the applied DC magnetic field. The nuclear magnetic moments of Rb atoms are transferred to ^{129}Xe and ^{83}Kr atoms according to the so called spin exchange process [5]. Larmor precession of ^{129}Xe and ^{83}Kr spin axes starts when the two AC magnetic fields are applied along the two directions orthogonal to the direction of the DC magnetic field. This precession is observed by exploiting the component along x-axis of optical signal illuminating the NMR cell according with the technique developed in [6].

Recently a K–^{3}He NMR gyro exploiting optical excitation of Larmor precession and optical read-out of the rotation-induced NMR frequency shift has been experimentally demonstrated in [7]. This gyro exhibits a resolution in the range 0.01–0.1 °/h. This resolution is very similar to those of the He–Ne RLG and the IFOG.

The NMR gyro is a very complex and expensive apparatus and its miniaturization seems to be quite difficult. At this stage it does not seem competitive with competing technologies in the high performance gyro market.

7.2.2 Gyro Based on Atom Interferometer

Sagnac effect is valid not only for photons but also for other massive particles such as atoms, neutrons and electrons. An innovative gyroscope exploiting wave packets of caesium atoms has been reported in [8]. The operating principle of this angular rate sensor is similar to that of a phase sensitive optical gyroscope.

The phase shift $\Delta\varphi_a$ between two atomic beams counter-propagating in a rotating interferometer is given by:

$$\Delta\varphi_a = 2\frac{m}{\hbar}\left(\overrightarrow{\Omega} \cdot \overrightarrow{A}\right) \tag{7.2}$$

where $\hbar$ is the reduced Planck constant, m is the particle mass, $\overrightarrow{\Omega}$ is the interferometer angular rate and $\overrightarrow{A}$ is the area spanned by the counter-propagating beams. If we compare atom and optical interferometers having the same area, the rotation induced phase shift experienced by the atomic beams in an atom interferometer is very larger than that experienced by photon beams in an optical interferometer. This justifies the research interest in angular rate sensors based on counter-propagating atomic beams.

In the gyro proposed in [8] two laser-cooled beams of cesium atoms propagate in a squared interferometer. Beams in the interferometer are divided, deflected and recombined by two-photon stimulated Raman transitions.

Device resolution is in the range 0.001–0.01 °/h, which is better than that assured by the IFOG and He–Ne RLG.

A cold atom rotation sensor based on the same operating principle has been very recently reported [9]. Resolution less than 10^{-3} °/h has been predicted for this sensor.

Gyroscope based on the atom interferometers is very complex and expensive and so its usage is justified only for specific applications, such as the experimental verification of general relativistic effects, requiring outstanding performance.

7.2.3 Superfluid Gyroscope

Superfluids are fluids that do not exhibit any viscosity. A typical example of superfluid is the ^{4}He that exists as a superfluid at temperatures below 2.17 K.

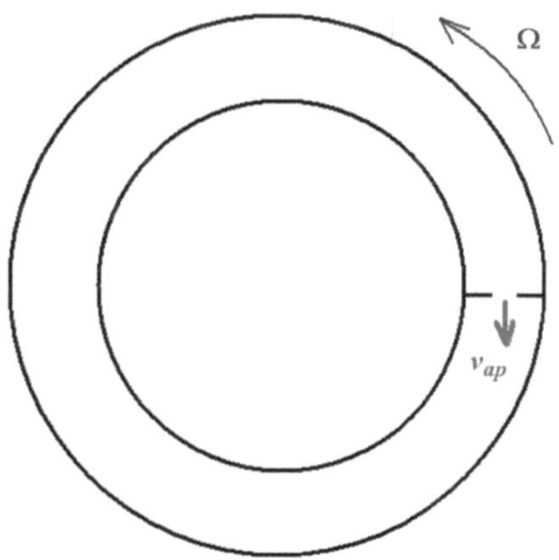

Fig. 7.3 General scheme of a superfluid gyroscope

Superfluid gyro consists of a torus partitioned by a wall containing an aperture having an effective width equal to l (see Fig. 7.3). The torus is filled by a superfluid and when it rotates a backflow through the aperture is induced. The superfluid velocity in the aperture is equal to:

$$v_{ap} = -2\frac{\overrightarrow{\Omega} \cdot \overrightarrow{A_t}}{l} \tag{7.3}$$

where $\overrightarrow{A_t}$ is the area spanned by torus arms and $\overrightarrow{\Omega}$ is the torus angular rate. By monitoring superfluid velocity in the aperture it is possible to estimate torus angular rate.

In [10] a multiturn superfluid gyro having an area of 95 cm^2 is reported. The exploited superfluid is the ^{4}He at temperature of 0.28 K. Experimentally demonstrated sensitivity of this gyro is about 1 °/h. Main advantage of this sensor is its long-term stability.

References

1. Canet, D.: Nuclear Magnetic Resonance: Concepts and Methods. Wiley Interscience, Chichester (1996)
2. Woodman, K.F., Franks, P.W., Richards, M.D.: The nuclear magnetic resonance gyroscope—a review. J. Navig. **40**, 366–384 (1987)
3. Grover, B.C., Kanegsberg, E., Mark, J.G., Meyer, R.L.: Nuclear magnetic resonance gyro. US Patent 4,157,497, 1979
4. Greenwood, J.A.: Nuclear gyroscope with unequal fields. US Patent 4,147,974, 1979
5. Bouchiat, M.A., Carver, T.R., Varnum, C.M.: Nuclear polarization in He^3 gas induced by optical pumping and dipolar exchange. Phys. Rev. Lett. **5**, 373–375 (1960)
6. Cohen-Tannoudji, C., Dupont-Roc, J., Haroche, S., Laloë, F.: Diverses résonances de croisement de niveaux sur des atomes pompés optiquement en champ nul. Rev. Phys. Appl. **5**, 95–101 (1970)

7. Kornack, T.W., Ghosh, R.K., Romalis, M.V.: Nuclear spin gyroscope based on an atomic comagnetometer. Phys. Rev. Lett. **95**, 230801 (2005)
8. Gustavson, T.L., Landragin, A., Kasevich, M.A.: Rotation sensing with a dual atom-interferometer Sagnac gyroscope. Class. Quantum Grav. **17**, 2385–2398 (2000)
9. Müller, T., Gilowski, M., Zaiser, M., Berg, P., Schubert, Ch., Wendrich, T., Ertmer, W., Rasel, E.M.: A compact dual atom interferometer gyroscope based on laser-cooled rubidium. Eur. Phys. J. D **53**, 273–281 (2009)
10. Bruckner, N., Packard, R.: Large area multiturn superfluid phase slip gyroscope. J. Appl. Phys. **93**, 1798–1805 (2003)

Index

 MIX
Papier aus verantwortungsvollen Quellen
Paper from responsible sources
FSC® C105338

If you have any concerns about our products,
you can contact us on
ProductSafety@springernature.com

In case Publisher is established outside the EU,
the EU authorized representative is:
Springer Nature Customer Service Center GmbH
Europaplatz 3, 69115 Heidelberg, Germany

Printed by Libri Plureos GmbH
in Hamburg, Germany